ALSO BY BOB McDONALD

Canadian Spacewalkers:
Hadfield, MacLean and Williams Remember
the Ultimate High Adventure

Measuring the Earth with a Stick:
Science As I've Seen It

Wonderstruck II

Wonderstruck

Everything You Ever Wanted to Know About

Black Holes, Dwarf Planets, Aliens, and More

An Earthling's Guide to OUTER SPACE

BOB McDONALD

PUBLISHED BY SIMON & SCHUSTER
New York London Toronto Sydney New Delhi

Simon & Schuster Canada
A Division of Simon & Schuster, Inc.
166 King Street East, Suite 300
Toronto, Ontario M5A 1J3

This Simon & Schuster Canada edition October 2019

SIMON & SCHUSTER CANADA and colophon are trademarks of Simon & Schuster, Inc.

For information about special discounts for bulk purchases, please contact Simon & Schuster Special Sales at 1-800-268-3216 or CustomerService@simonandschuster.ca.

Interior design by Lewelin Polanco
Illustrations by Tony Hanyk, tonyhanyk.com

Manufactured in Canada

10 9 8 7 6 5 4 3 2 1

Library and Archives Canada Cataloguing in Publication
Title: An earthling's guide to outer space : everything you ever wanted to know about black holes, dwarf planets, aliens, and more / Bob McDonald.
Names: McDonald, Bob, 1951– author.
Identifiers: Canadiana (print) 20190094664 ffl Canadiana (ebook) 20190094672 ffl ISBN 9781982106850
(hardcover) ffl ISBN 9781982106867 (ebook)
Subjects: LCSH: Outer space—Miscellanea—Juvenile literature. ffl LCSH: Universe—Miscellanea—Juvenile literature. ffl LCSH: Astronomy—Miscellanea—Juvenile literature.
Classification: LCC QB500.22 M34 2019 ffl DDC j520.2—dc23

ISBN 978-1-9821-0685-0
ISBN 978-1-9821-0686-7 (ebook)

To curious minds everywhere

CONTENTS

PART 1

The Great Beyond

Answers to the Big Questions

1

How Big Is Our Galaxy?

Our Earth is floating through a beautiful whirlpool of stars almost too large to imagine. We call our galaxy the Milky Way because it looks like someone spilled milk across the night sky. You can see it with your own eyes, but it's a little tricky. You have to find a dark place where there are no streetlights, away from towns and cities, on a clear night when the moon is not up. If you can find a spot like that—in the country, beside a lake, on a farm—and look up on a summer night, you'll see a ghostly glow arcing across the entire sky, a bridge of stars that reaches from horizon to horizon.

That's our home galaxy. But there's much more to it than meets the eye. Looking up from the ground at night, we don't see the whole galaxy because we're inside it, the same way you can't see your entire town or city from your front door.

If you could soar above the Milky Way, what a spectacular sight you would witness—a luminous swirl of stars with four graceful, curving arms wrapping around one another in a pinwheel shape. Our sun is just one of hundreds of billions of stars on that merry-go-round. We live on the inside edge of one of the galaxy's curving arms, about two-thirds of the way out from the center, and the nearest star to our sun is more than four years away at the speed of light.

If you wanted to travel around the entire galaxy the way fictional starships in movies do, you'd need a fast ship—one that could move at the speed of light, at the very least. And that is crazy fast. At light speed, you'd cover three hundred thousand kilometers, or seven times around the Earth,

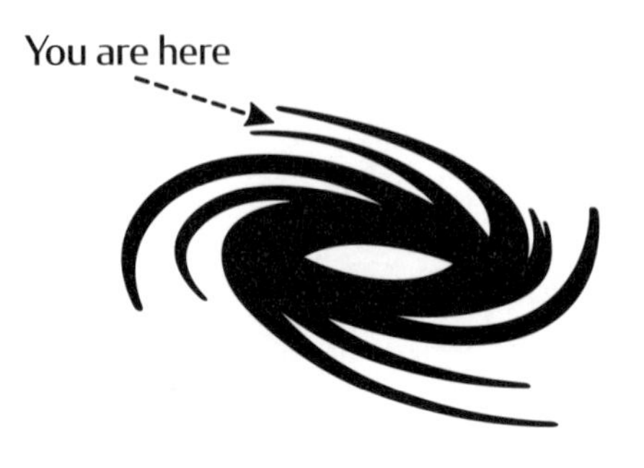

in one second. You could visit the moon and come home again in a second and a half, or travel to Mars in twenty minutes.

But even traveling at that incredible speed, it would still take you thirty thousand years to travel from Earth to the center of the galaxy. And if you tried to cross from one side to the other, it would take one hundred thousand years.

SPACE PLACES

Take a walk through the Milky Way Galaxy in the Galaxy Garden in Hawaii. This big, circular garden was designed in the exact shape of the Milky Way, with different plants representing the stars and nebulae. A funnel-shaped fountain in the middle of the garden represents the black hole in the center of the galaxy, and all of the plants are arranged in curving arms that resemble the spiral shape of our galaxy.

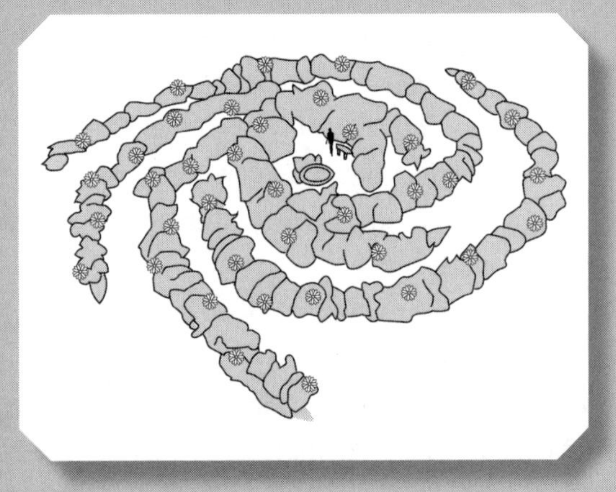

If you want to find the Earth, you will have to look hard, because on the scale of this garden, it is too small to see with the eye. But there is a spot, a little more than halfway out from the center fountain, where a little dot on the leaf of one plant represents our sun.

If you could make the journey to the center of the galaxy, you'd see it for what it truly is—a city of stars. About 300,000,000,000, or three hundred billion stars.

That number is so huge, it's hard to wrap your head around. But you can get a sense of how many that is by imagining every star as a grain of sand. If you pick up a handful of sand, you're holding a few thousand grains. That's about the number of stars you can see with your eyes on a clear night. But that's only a tiny fraction of what's really out there.

An average dump truck holds approximately as many grains of sand as there are stars in the Milky Way. Imagine that dump truck full of sand spread out roughly thirty meters across, or the area of a baseball diamond. That's the size of our galaxy. But it's only one of about a hundred thousand million other galaxies that make up the rest of the known universe.

So how many stars are there in the universe? At least this many: 200,000,000,000,000,000,000,000. That number is so big, it doesn't even have a name.

If all the stars in the universe were turned into grains of sand, they would cover all the beaches in the world.

After you've pondered that, pick up just one little grain of sand on the end of your finger and look at it. That's one star—our sun. At that scale, the Earth is not even visible to your eye.

The universe is incredibly vast and filled with an unimaginable number of stars. It can make you feel small and insignificant when you think about it all. But remember this: we may be small, but at least we know our place in the universe!

Of course, for now, we don't have any spaceships that can take us across the galaxy, so we have to journey there in our imaginations. To do so, lie on the ground. Don't think about the sky as "up." Think of it as a huge space that reaches out in all directions. Remember, there are just as many stars "below" you as there are above, because you're lying on a big ball. And you are not on top of the ball—that's the North Pole. You're somewhere on the side. (The next time you see a globe of the Earth, find your city and see what side you live on.) If you think of yourself like a fly on a wall, with the ground behind you instead of below you, and the stars in front of you instead of above, you'll get a truer sense of your place in space.

THE FUTURE IS NOW

It doesn't happen very often, but every once in a while, two galaxies run into each other. Don't worry, no one gets hurt. Galaxies are mostly made of empty space, with the stars so far apart that two galaxies can pass right through each other without anything actually touching. But when they get too close, gravity can pull them out of shape, rearranging those long spiral arms.

Eventually, our Milky Way is going to collide with the next-biggest galaxy close to us, Andromeda. Luckily, that won't happen for another four billion years! When it does, though, there will be two Milky Ways in the sky!

YOU TRY IT!

Model Galaxy

WHAT YOU NEED

1. A large bowl
2. Water
3. Food coloring or milk

WHAT TO DO

1. Fill the bowl with water.
2. Using your finger, swirl the water around in a circle until it dips down a little in the center.
3. Let the water spin on its own for five seconds, then carefully pour just one or two drops of food coloring or milk in the center.
4. Watch the shape the food coloring or milk takes.
5. You have just made a model of the Milky Way!

The food coloring or milk should take on a spiral shape, with long arms wrapped around the center of the bowl. That is the shape of our Milky Way Galaxy. The sun and billions of other stars all swarm around a central bulge. Galaxies come in many different shapes, but the spirals, like ours, are the most beautiful. Don't you agree?

2

Is There Life in Space?

With so many stars and planets in the universe, it seems silly to think that we're living on the only planet with life. So if there are alien life-forms out there, where are they? No other life has been found on worlds other than Earth, but that doesn't mean it's not out there.

Consider how many different forms of life there are on our planet. Plants and animals have found ways to live in extreme environments—not only in lush jungles and forests but also in hot, dry deserts, in frigid ice, and even in boiling water at the bottom of the ocean. Living things that survive in extreme environments are called "extremophiles." And if extremophiles can exist in the most challenging landscapes on Earth, what about in

weird environments on other worlds? Could there be life in the incredibly cold, dry deserts of Mars? Might there be creatures floating around in the orange clouds of Jupiter? Or strange life-forms hiding under the ice of distant moons?

We don't know the answers to these questions yet, but one thing we do know is that when we find life on other worlds, it probably won't be the little people with big heads and huge eyes we see in the movies. Those movie aliens are often large creatures, too, but when we do find life on other worlds, there is a good chance it will be very small. Higher forms of life, like animals, fish, and birds, take billions of years to evolve. We will likely find microscopic cells floating in water, simple plants, or pond scum. That's what life on Earth was like for most of its history, and even today, there are far more insects and microscopic creatures than there are people or elephants.

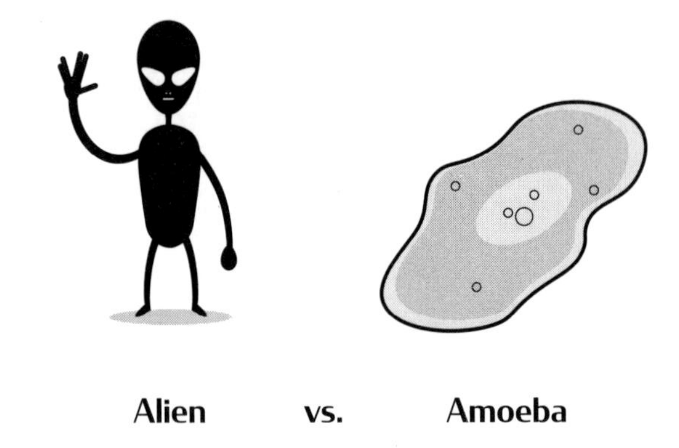
Alien vs. Amoeba

Another thing to consider is that all life you see around you—whether plant, animal, bird, or insect—is made of the same basic ingredient as you are: carbon. Some scientists have suggested that alien life could be based on another element, such as silicon, which makes up rocks and sand. Can you imagine living rocks? We might not even recognize them as alive. Alien life may be so strange-looking to us we might have a hard time relating to it.

So where do we start our search for life on other worlds?

We've begun our search of Mars, using robots. Mars is cold and dry today, but scientists believe there used to be lakes, rivers, and even an ocean on its surface about three billion years ago, roughly the same time that life was just starting to appear on planet Earth. Once upon a time, you could have gone swimming and sailing on Mars! Was there life in those

Martian oceans? That's one of the questions our robots are trying to answer. They are searching for signs of life, but so far, they haven't turned up anything.

Today, Mars remains covered with markings that look like rivers and lakes. The only problem is that we don't see any water in them. It seems they dried up when Mars went into an ice age. Today, it's hard to find a place on Mars that isn't below freezing, which decreases the chances of finding alien life-forms.

But maybe we're looking in the wrong places. What if there's life *in* Mars rather than on it? There may be caves where it's warmer and wetter, caverns where extremophiles could survive. The problem is, you have to know where the entrances to Martian caves are. So far, our robotic rovers haven't had much luck in finding any. And even if we do find caves on Mars, there's another problem. The air on Mars is different from the air on Earth. Martian air doesn't have oxygen in it, so we won't recognize any life we find there.

One thing we see on Mars is lots of ice. Ice is everywhere in space! And we already know that even in the coldest places on Earth, microscopic life can survive. If life can exist underneath ice on Earth, perhaps we could find life under the ice on Mars. All we need is some way of digging down into that frozen water to find it.

Some scientists think there's a better chance of finding life on Europa, one of Jupiter's moons, than on Mars. Europa is completely covered in ice, and under that ice is an ocean containing more salt water than all of the oceans on Earth. There could be hot-water vents at the bottom similar to those on our planet, with their own kind of microscopic extremophile life.

Another way we might find out if there's life in space is through space probes sent to comets. When the satellite *Stardust* flew to a comet in 2004, it scooped up some material from the comet's tail, then returned to Earth. The comet dust it retrieved contains the same basic chemicals—called organics—that life is made of, which begs the question: Is it possible that life on Earth came from comets?

This concept has a name: "panspermia," meaning "life that spreads around the universe via comets." Microscopic organisms (or even the

chemicals that make up life) buried deep inside an icy comet could be protected from the harsh environment of space for millions of years. Then, when a comet strikes a planet such as Earth, which is warm and has lots of water, those life-giving ingredients may become the seeds for a whole generation of creatures. We know the Earth has been hit by comets many times in the distant past. What does this mean? Maybe we're all descendants of alien matter from space!

If you're old enough, you may recall ads for sea monkeys in the backs of comic books. News flash: there's no such thing as a sea monkey. These creatures are actually tiny shrimp, and they have an amazing ability to survive when their ponds dry up. They can dry out so much that they look like dirt and appear dead. They can stay in this state of suspended animation for many, many years. To bring them fully back to life, all you need to do is put them in water.

It's astounding how something that seemed dead can come back to life. This kind of hibernation could allow life to travel through space hidden inside a comet. How many seeds of life are floating around out there right now, just waiting to run into a warm, wet planet like ours? When we search for life on other worlds, this is the type of creature we're likely to run into first. They're life-forms that, in many ways, are more resilient than us.

Other planets may have water, but they are frozen planets. Venus, on the other hand, is a hot planet where water, if there is any, would exist only as a gas. The Earth is right in the middle. It's sort of like Goldilocks: not too

ON THE DRAWING BOARD

Is there anyone out there we can talk to? With all the billions of stars in billions of galaxies in the universe, it seems unlikely we are the only intelligent life. One way to search for other communicative civilizations is to listen for radio signals. We've been keeping an ear out for aliens for some time now. A number of radio telescopes use a process called SETI—Search for Extraterrestrial Intelligence—to listen for alien radio signals. So far, nothing has turned up.

If you want to join the search for aliens, you can help through a project called SETI@home. There are so many signals coming from space that scientists can't analyze them all. Through SETI@home, you can link your computer to a worldwide network that looks for alien signals in the data gathered by telescopes. Maybe you will be the first to locate a signal from an alien civilization!

hot, not too cold, but just right. Astronomers call this midrange "the habitable zone" or the "Goldilocks zone." In our solar system, this zone extends inward just a little toward Venus and outward to Mars. Astronomers and scientists continue to look for planets that might be habitable like ours somewhere in the galaxy. We have found a few, but no signs of life . . . yet.

YOU TRY IT!

Alien Linguistics

What would you say to an alien if you had the chance?

WHAT YOU NEED

1. Blank paper
2. Markers, pens, or colored pencils
3. A few friends

WHAT TO DO

1. Divide yourselves into pairs. Each pair should have one piece of paper and a couple of writing utensils.
2. Separate yourselves so you can't see what the others are doing.
3. Come up with a message that you would like to send to an alien civilization. It could be a greeting that says "Hello." Or perhaps you are in trouble and need to ask for help. Or maybe you want to invite them to come visit. Be creative, and don't tell the other group what you're trying to say.
4. Write your message on the paper for the other group. One rule: you cannot use English words, numbers, or any symbols that we use on Earth because the aliens will not know our language. It has to be some kind of drawing that gets your message across.

When you have completed your message, gather everyone back together and see if the other pairs can understand your message. Can you understand what they were trying to say to you? It is harder than you think!

3

Where Do Stars Come From?

Stars come from clouds. Just as clouds in the sky make raindrops, enormous clouds in space give birth to stars.

Space clouds are called nebulae. They look like beautiful space flowers drifting among the stars. Some are so large, it would take longer than your entire life to cross from one side to the other, even if you were moving at the speed of light. Nebulae are mostly made of hydrogen gas, which is very light, and they are found everywhere in the universe. It makes sense, then, that most stars, including our sun, are made almost entirely of hydrogen, too.

But while these nebulae are beautiful to look at through a telescope, if you were to pass through one on a spaceship, they sure wouldn't smell like flowers. They're full of dust and other carbon-based gases that are similar to what comes out of a car's tailpipe. So nebulae would probably smell like a combination of vehicle exhaust, smoke, and burned hamburger. The plus side? That space pollution is the fertilizer for new solar systems.

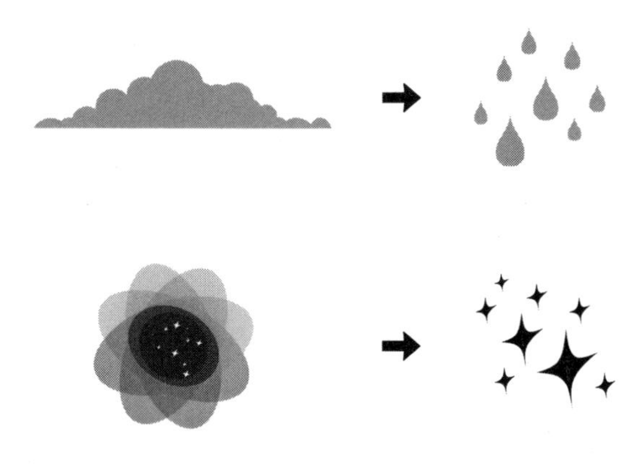

Clouds in space usually have some parts that are thicker than others. And in space, whenever anything—whether it's gas or rocks—gathers together in clumps, these clumps have more gravity than the rest of the material around them. The clumps within a space cloud attract more gas and dust, and as they do, their gravity grows, squeezing everything toward the center until a ball is formed. If that ball gets big enough, the center becomes extremely hot. Nuclear fires ignite . . . and a star is born.

Meanwhile, other smaller clumps of matter—the ones that don't fall into the center and become the star itself—circle around the new star, forming little balls that eventually grow to become planets. This is how the Earth and all the planets in our solar system were formed four and a half billion years ago.

You might be wondering, "Where did those big clouds in space come from in the first place?"

It turns out that many nebulae come from old stars that died a long time ago. Stars are born, but they also die. And when they do, their remains become new stars.

Stars can live for billions of years. (Ours, the sun, is 4.5 billion years old, but only halfway through its life.) Eventually, every star runs out of gas. The nuclear fires inside stop pushing outward, and gravity crushes the star down, squeezing it into a smaller space. The squeezing makes the center of the star even hotter, and for a short while, a different kind of burning at the center of the star begins pushing back out again. This puffs the star out into a big, red giant.

When our sun reaches this stage of its life, that will be the end of the Earth and all life on it. The sun will completely fill our skies. Temperatures on Earth will quickly rise until the oceans boil into steam, forming thick clouds around the whole planet. The heat of the big sun will burn these clouds off into space, giving the Earth a long, white tail like a comet. As the hot gases from the swollen sun surround the Earth, mountains will melt into valleys until the whole planet turns into a molten ball of boiling rock that will eventually become a burned-out cinder inside a dying star.

But let's look at the upside: with one final dying gasp, the sun will blow off its outer layers, along with the debris from what was our solar system, all of which will drift off into space, forming a brand-new nebula. From the death of stars comes the life of new stars and planets.

If you've ever wondered what happens to stars when they stop shining, touch your face. We are made of stardust. Today we live, but one day we will die. The atoms in our body will be incorporated into the ground, and then, a long time from now, the Earth and everything on it will be vaporized and cast out into space, where parts of it will reform into beautiful nebulae.

Who knows? Perhaps one day, way in the future, the atoms that make up your body will become part of a planet going around a new star. Maybe life will evolve on that planet and one of those life-forms will look up into the night sky and wonder where it all came from.

ON THE DRAWING BOARD

Here's a wild idea: we might someday build a spaceship called the interstellar ramjet that uses clouds in space as rocket fuel. It would be a very large ship sent far away to other star systems. If the journey involved passing through a giant hydrogen cloud, the ship would extend an enormous invisible scoop made of a magnetic field. The field would gather up some of the gas floating in space and funnel it into the ship's engines. There, the hydrogen would be burned and blown out the back end of the ship, making it go faster. A ship like this could theoretically go faster than any vehicle ever made, cutting down the travel time to the stars. There's just one little problem we still have to solve: Once you get a ship this large traveling that fast, how do you stop it?

To a faraway alien astronomer four billion years from now, our sun's nebula will probably look like a ring or a shell. That's because our sun is a fairly average-sized star. But there are giant stars out there, much bigger than our sun, and when they die, they go out with a bang called a supernova.

Supernovas are so violent and hot, they act like giant blast furnaces, forging heavy elements like nickel and gold. Think about that. When you see a gold ring or any piece of gold jewelry, you're looking at an element that was made not by the Earth but inside a dying star during a supernova explosion billions of years ago.

You can see an awe-inspiring nebula with your own eyes. All you need to do is find the constellation Orion. Orion comes around every winter and is the most easily recognized star pattern after the Big Dipper. The three stars in a row are the belt. Hanging from it is a sword marked by three more stars pointing downward. With binoculars, you'll see that the middle point of the sword looks a bit fuzzy. The misty part isn't actually a star but a nebula. This gigantic cloud, almost 1,500 light-years from Earth, is a cosmic nursery containing hundreds of stars still in the process of forming, plus other stars already born. Four of them, together known as the Trapezium, are baby suns that flared to life less than a million years ago. The intensely hot Trapezium stars are what make the Orion Nebula glow.

4

What Is Dark Matter?

Space seems to be mostly empty and dark. At least, that's how we thought about it for a long time. But now astronomers think something is hiding in the darkness. It's everywhere, it's powerful, and no one knows what it is.

They call it dark matter. And when scientists say "dark," that really means, "We don't know, but we're working on it."

The first clue that there's something out there in the darkness came from watching how things move in space. Almost everything in space moves in circles. Planets go around stars. Stars clump together and swirl around in galaxies, and galaxy clusters spin around one another like bugs flying around a light.

About eighty years ago, though, an astronomer named Fritz Zwicky discovered an anomaly. He was studying a group of about a thousand galaxies known as the Coma Cluster. He wanted to know just how much gravity it took to hold that particular cluster together. Gravity comes from mass, and when Dr. Zwicky calculated the mass of all the galaxies, something didn't add up. The galaxies simply didn't have enough mass by themselves to produce enough gravity to hold the group together. Those galaxies should have been flying away from one another!

So what was holding them together? Dr. Zwicky figured there had to be something else in that cluster that he couldn't see, some sort of invisible mass that was producing extra gravity. And the astounding part was that there seemed to be more of that mystery material in the galaxies than there were stars.

Scientists aren't very good at naming things. Whatever was holding everything together was invisible—or dark. And since it was something, it might be matter. So "dark matter" is what it was called.

But how do you study something invisible? Astronomers have found galaxies in space that have been bent out of shape or stretched into long, thin arcs that look like eyebrows. The light from those distant galaxies is distorted by invisible dark matter that lies between us and those galaxies. So even though dark matter is invisible, it gives itself away by distorting light.

Scientists think dark matter could be made of invisible particles that are so small they pass right through the Earth, the same way light (which is also made of waves of particles) goes through a window. Scientists have found some invisible particles in space called neutrinos. They come from the sun and other stars in the galaxy. But there are not enough neutrinos in the universe for them alone to make up all dark matter. There must be something else.

We can study neutrinos, but it's a tricky process. To capture something that travels right through the Earth, you have to go *inside* the Earth, which isn't easy. Deep underground in northern Ontario is the Sudbury Neutrino Observatory. It uses a big ball filled with a special clear liquid, called heavy water, that is used inside nuclear reactors. The atoms in this water give off a tiny flash of light if a neutrino hits them. The observatory saw enough flashes to prove that neutrinos exist. Many neutrinos come from the sun, which means they shine down on your head during the day. But since they pass right through the Earth, they also shine up under your feet at night! Now scientists in the same underground lab can further study the nature of these ghostly particles and are starting to look for others that might be dark matter.

Some scientists think that the answer lies in massive particles that are invisible but not very strong. They call them WIMPs—Weakly Interacting Massive Particles—because they don't bother interacting with ordinary matter like the stuff you and I are made of.

When you drop a ball on a table, it bounces. That's an example of interaction. The ball and the table push against each other. But if you drop a WIMP on the table, it won't bounce. It will pass right through the table. Then it will continue through the floor, through the Earth, and out the other side into space. So even though a WIMP has more mass than a neutrino, it passes through solid matter as if it weren't there. Remember, this is just a theory. Nobody has actually spotted one of these strange subatomic particles—yet.

If all the dark matter we're looking for is made of WIMPs, this answers the mystery of what keeps the universe in balance. But if there's something else out there, we have yet to find it.

Consider this: when you're outside on a clear, starry night, the blackness you see is what most of the universe is made of. No one knows exactly what that dark stuff is, but it is believed to make up 95 percent of the universe. What constitutes the rest? Galaxies, stars, planets—and us.

5

Why Is the Sun So Hot?

The sun is hot because it is incredibly big. If the sun were a beach ball, the Earth, by comparison, would be the size of a pea.

The sun weighs a billion, billion, billion tons—two quintillion kilograms (yes, that's a real number). When an object is that big, even if it's made entirely of gas, its weight squeezes everything together incredibly tightly. And when matter is squeezed together under such enormous pressure, it gets hot—up to fifteen million degrees Celsius, in fact. That's plenty hot enough for the sun's gases to turn into a hot plasma.

A plasma is a glowing, electrified gas, like the white hot flash of a lightning bolt. The flame in fire or the glow in neon lights are other examples. But none of these are as hot as the plasma at the center of the sun. Atoms in the sun's plasma fly around in all directions, sometimes smashing into one another at high speed and fusing together to make bigger atoms. This is similar to the way two water drops might join together in a rain cloud to become a larger one. In the case of plasma, this

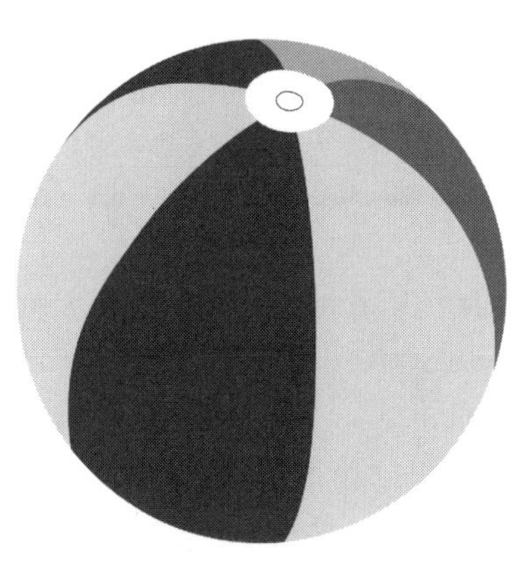

process of fusion gives off a lot of heat and light, which is the energy that makes the sun shine.

ON THE DRAWING BOARD

Scientists on Earth would like to duplicate the fusion happening in the sun so everyone in the future has lots of clean energy. But replicating the process is hard to do because when you heat up plasma to fifteen million degrees, it melts whatever container you put it in. We've managed to create fusion, but only for a second or so inside a nuclear bomb, which is the most powerful bomb there is.

There is a less destructive way to duplicate the sun on Earth, though. The world's largest fusion reactor is under construction in France. It's called ITER, which is Latin for "The Way." It plans to use a donut-shaped chamber to heat a form of hydrogen up to 150 million degrees Celsius, ten times hotter than the center of the sun. All that heat from the fusion will warm up liquids in the walls of the reactor, until that liquid turns to steam that will be used to run generators and make electricity.

The biggest challenge with ITER is that it will require a huge amount of energy to heat the hydrogen plasma up to that temperature. And it will take still more energy to run the powerful magnets needed to keep the plasma floating in the center of the donut without actually touching the walls. Scientists hope that the fusion reactions will eventually sustain themselves the way they do in the center of the sun, producing more energy than what is needed to run the machine. If they succeed, then fusion reactors may provide clean energy for the future.

In a sense, the sun is a big bomb that never stops exploding. So why doesn't the sun blow itself to pieces?

Because it's big.

There is a battle between two forces: the endless explosions in the center pushing out and gravity pulling everything back in. Neither force wins.

The two balance each other out and the sun continues to shine brightly in our skies.

In case you're worried that the sun will run out of fuel, know that the sun has so much gas that it will keep burning for at least another four billion years. So don't worry: it will be there at sunrise tomorrow morning.

THE FUTURE IS NOW

Out in space, when sunlight reflects off an object, it pushes. The push is very tiny, and you wouldn't feel it on your body, but if you had a mirror large and light enough, you could capture the pressure to sail around the solar system, all for free! The biggest challenge would be building a sail large enough to harness the solar energy. The sail would have to be at least a kilometer across to capture enough energy. Making a sail that large work

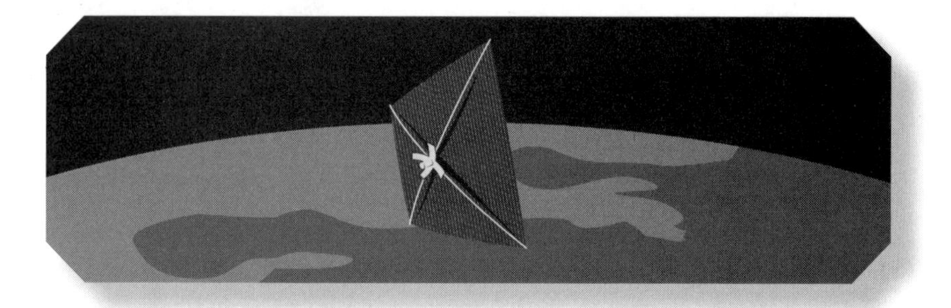

on Earth would be impossible, but in space, everything is weightless. You could make the sail out of thin, shiny plastic—similar to the plastic used in potato chip bags—fold it into a small package, and launch it into space on a rocket. Then it would unfurl like a flower and take your ship to the moon or Mars purely on the power of sunlight. Imagine a future when solar sailors ride sunbeams to far-off planets.

If you could somehow get yourself to the surface of the sun and not be burned up by the scorching temperatures (the sun is cooler on the outside—only 6,000 degrees Celsius), you would not be able to stand up

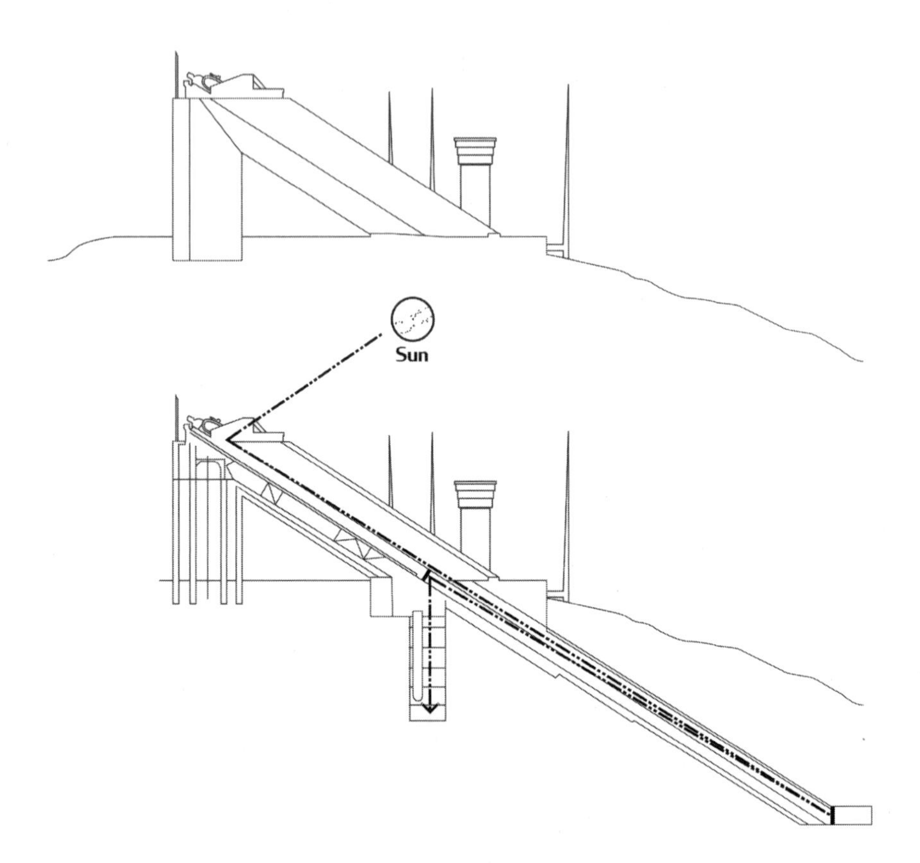

because the sun's gravity is twenty-eight times stronger than what we feel on Earth. A person weighing sixty kilograms on Earth would weigh 1,680 kilograms on the sun.

I hope this convinces you not to go walking on the sun any time soon. You can always admire the sun by looking at it—just not with the naked eye. Scientists examine the sun using solar telescopes, which have special filters that stop most of the sunlight from getting into the sensitive instruments. One kind of solar telescope—the world's largest, in fact—uses only a mirror.

The McMath-Pierce telescope in Arizona reflects sunlight down a long tunnel that's filled with spectrographs—which analyze the sunlight—and mirrors, which capture the image of the sun's surface on a big white screen near the bottom. The image of the sun on that screen is still so bright that scientists viewing it need to wear special goggles.

So what is it they see? The sun's face is covered with black spots called sunspots. Most of those spots are larger than the entire Earth and are evidence of the violent activity taking place inside the sun. Scientists watch as the spots grow, hang around for a while, then disappear, only to return again in the course of an eleven-year cycle.

Scientists believe sunspots are formed when different parts of the sun swirl around one another. Because the sun is made of gas, it rotates faster at the equator than it does at the poles. That difference causes the magnetic field of the sun to twist. In some areas, the magnetic field breaks through the surface, reaches into space, and curves back down again. Enormous arcs of hot gas, so big you could pass the Earth underneath them, follow these magnetic lines, connecting the sunspots, which lie at the ends of the arch like pots of gold at the ends of a rainbow.

Sometimes these arches grow so big and tall that they snap, sending gigantic blobs of the sun's plasma into space at more than one million kilometers per hour. Those projectiles are called coronal mass ejections, and every once in a while one of these blobs heads toward the Earth. They've been known to knock out satellites and cause power failures.

If an eruption happens on the sun, scientists send a warning so that companies with sensitive electrical systems can get ready. It's a new kind of space weather forecasting.

We don't have to fear these eruptions much. While they can affect our technology, they don't hurt people because our atmosphere protects us. In fact, we get to see the effects in that beautiful light show called the aurora borealis (or aurora australis in Australia). The particles from the sun hit our air and make it glow the way electricity lights up the gases in neon signs. So if you're fortunate enough to see the northern lights, you're actually seeing a rain of light from space. You're seeing a part of the sun hitting the Earth!

YOU TRY IT!

Solar Telescope

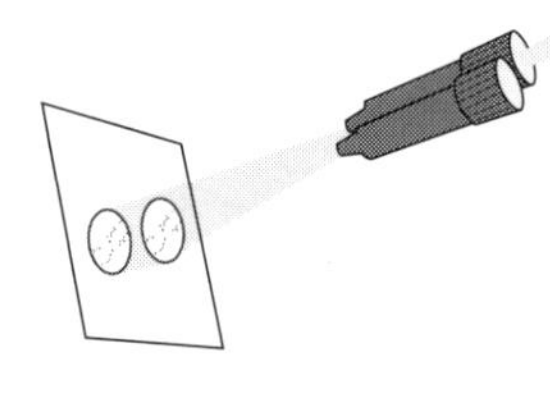

WHAT YOU NEED

1. One pair of binoculars
2. One sheet of white paper

WHAT TO DO

1. Turn your back to the sun. (NEVER LOOK AT THE SUN THROUGH BINOCULARS! YOU'LL BE BLINDED INSTANTLY.)
2. Hold the binoculars on your shoulder with the large end facing the sun.
3. Hold the paper in front of you until you see the shadow of the binoculars on the paper.
4. Move the binoculars around until two bright spots appear on the paper. While holding them as steady as you can, turn the focus on the binoculars until the circles become sharp. You should see two images of the sun. Can you see any little black spots?
5. You'll probably see the image shake a lot, which is actually the movement of your body. You can make the image steady by propping the binoculars up on the edge of a table or a pile of rocks instead of your shoulder.
6. You are now looking at the face of the sun.

The little black dots will look small to your eye. In fact, they may be hard to see at all. But remember that you could drop the entire Earth into each one of those dots. The sun is unimaginably huge!

6

Do UFOs Really Exist?

Yes, they do!

They're objects moving in the night sky, and most times, you don't know what they're doing up there. They're unidentified flying objects—UFOs—and they're very real. But that doesn't mean they're alien spaceships.

Usually, we're able to figure out what an object in the sky is. What we might think is a UFO often turns out to be an airliner or a satellite or the International Space Station. Once a UFO is identified, it becomes an IFO—identified flying object.

IFOs are common, of course, but that doesn't stop many people from believing that what they glimpse is something else entirely. If all you see is a light in the sky, how do you know for sure what it is or isn't? The idea that alien spaceships could be visiting Earth is not that unreasonable. When you think about the thousands of millions of planets that could be out there

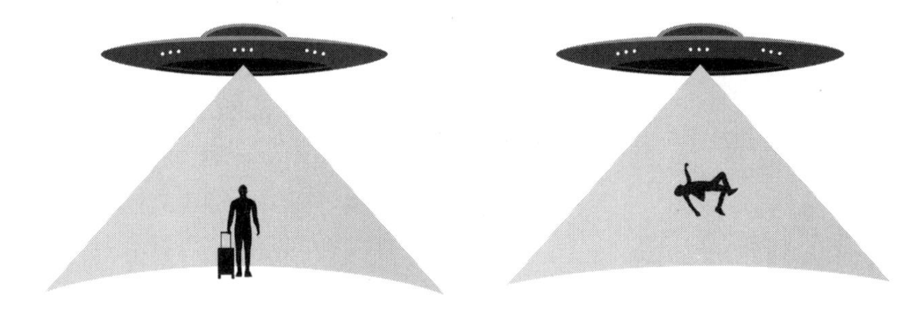

orbiting stars just like our sun, the chances are pretty good that some of those planets have life. And if some of that life is intelligent enough to build spaceships, maybe those life-forms would come to visit us. After all, we also strive to explore other worlds.

The idea of UFOs began in Roswell, New Mexico. In 1947, something was found about one hundred kilometers northwest of the city. Rancher Mac Brazel went to check his livestock following a massive thunderstorm. He came across an extensive area of debris the likes of which he'd never seen before. He gathered it all up and brought it to the sheriff's office. The sheriff didn't know what it was. Military gear, maybe? At the time, Roswell was home to the Roswell Army Airfield, the only atomic bomb base in the world. The sheriff called in one of the base's intelligence officers, Major Jesse Marcel, who gathered up all the debris, loaded it on an aircraft, and took it away to a secret base called Area 51.

On July 8, 1947, the military issued a press release saying it was in possession of a flying saucer. On July 9, General Roger Ramey issued another press release saying that the remains were not a flying saucer but a secret weather balloon that had landed unexpectedly.

That was the end of the story. Or was it? The public is not allowed into Area 51, and the military certainly doesn't want people poking around in there and asking questions. The military is involved in all kinds of secret projects. They're working on experimental new planes and vehicles they don't want anyone to see. They might fly them at night, which would lead sky watchers to say, "Hey, I just saw something really weird in the sky!" to which the military might reply, "It wasn't us." The military may cover up their own experimental airplanes, but could they and would they cover up an alien spaceship?

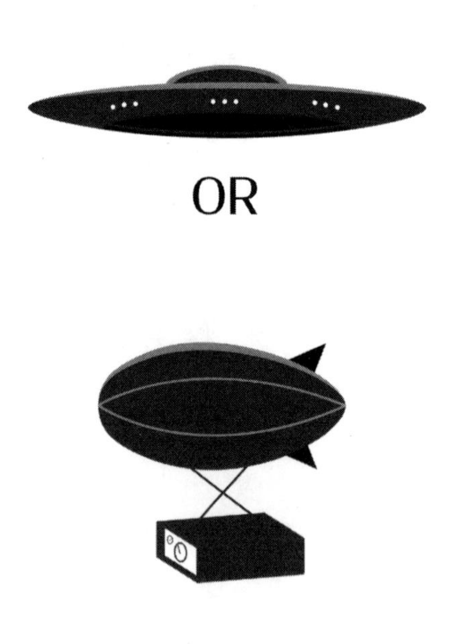

There are also no clear pictures

of alien spaceships flying through the skies. Photos that exist are always blurry, and videos are so shaky and out of focus that you can't tell what is being filmed. It's suspicious that no one has been able to take a proper photo of a UFO when we all carry phones that have cameras in them. If alien spaceships are flying around, why can't we just get some good, clear pictures? Maybe you will be the first to capture a good image of a real UFO.

SPACE PLACES

In the desert of Peru in South America, near the town of Nazca, are hundreds of long lines in the ground that run perfectly straight, sometimes for more than thirty kilometers. Some of those lines form giant drawings of animals, such as a hummingbird, a spider, and a monkey. The unusual part of these drawings is their size. They're so big, you can't see the shape of the animal they make from the ground. That picture can only be seen by flying over them in an airplane.

Most scientists—not to mention most of the public—agree that the lines were built by the Nazca people, who lived in the desert about two thousand years ago. But one small group of people who were not scientists believed these lines were made by ancient alien astronauts who came to Earth. This group claimed that the alien astronauts used the animal drawings as signs of where to land on Earth, claiming the long straight lines were landing strips. (Why aliens who were capable of traveling across vast distances of space would need landing strips was never addressed.) The argument was that because the ancient Nazcans had no flying machines, they could never make such large drawings from the ground alone.

Scientists who have studied the lines and drawings do not know exactly why they were drawn. Perhaps they're a giant calendar, pointing to constellations in the sky. Some of the lines do point to where certain stars appear on the horizon during the equinox. The animal figures can actually be seen from the ground if you climb nearby hills. Most scientists believe they were made for religious purposes, or built for the gods in the sky to see, especially the ones who bring water. The Nazca desert is one of the driest places on Earth, so water was a scarce and valuable resource to those people. The spider was a symbol of water, the hummingbird a symbol of fertility, and the monkey is associated with the Amazon jungle, which has a lot of water.

The Nazcans knew mathematics and how to build cities. They were talented at measuring distances and figuring out how to make something so large it could only be seen clearly from above. To this day, there is no credible evidence that their drawings were made by or for aliens who visited the Earth thousands of years ago.

Some people claim to have been kidnapped by aliens when they were asleep. They go so far as to say they had experiments done on their bodies before being returned to their beds. But so far, no one has been able to prove these claims.

Sleep paralysis may explain some so-called alien abductions. It's a condition that makes you feel like someone has taken control of your body. It's perfectly natural, and it happens all the time. When you go to sleep, the part of your brain that controls your muscles is partly shut down so your body will lie still. As you come out of sleep, the muscle control center in your brain is turned back on so you can move again. Once in a while, you can partly wake up during a dream while your control center is disconnected. It feels like paralysis because you can't move, even though you want to. It's as though someone or something else has taken control of your body.

The experience can be very scary—so scary, in fact, that in your dreamlike state you might believe aliens abducted you.

Beyond abduction reports, there have been many famous UFO hoaxes and elaborate practical jokes staged to fool the public. Some of them involve "alien" markings on the ground, such as crop circles. It turns out most of those were made by low-flying helicopters. Other hoaxes involve trick photography. Sometimes an object like a hat or a flying disc is thrown into the air and photographed in front of distant trees so that it looks like a UFO is flying above them.

What does all this amount to in our search for UFOs? Until we lay hands on an actual, verifiable piece of an alien spaceship or a real alien body, there's simply no proof that aliens have reached our world. For every sighting that could be an alien spaceship, there's almost always an alternate explanation—a secret military aircraft, a fireball meteor, a hoax, or a bad dream. And for those sightings that remain unexplained, there isn't enough evidence to say exactly what they are.

Speaking of the military, many organizations around the world are constantly watching space for unusual objects. They can see missiles being launched, satellites orbiting the Earth, or pieces of space junk as small as the size of your fist. At the same time, astronomers are scanning the skies looking for new comets or asteroids. It would be very hard, then, for an alien spaceship to approach Earth without a lot of people seeing it. And if one were spotted, why would it be covered up? It would be the news story of the century!

Let's not close our minds to the possibility of alien spaceships quite yet. We know the odds of intelligent life existing in the universe are pretty good. Some life-forms, like us, could develop spaceships—even faster spaceships than we have—and come to our planet. It's possible, just not provable . . . yet.

YOU TRY IT!

Make Your Own "UFO" Video

WHAT YOU NEED

1. A toy spaceship
2. Invisible thread
3. A camera
4. Proximity to a parking lot

WHAT TO DO

1. Set up the camera so you can see lots of cars in the parking lot.
2. Tie the toy spaceship to the end of the thread and hang it close to the camera so it looks huge and is hovering over the cars. Make sure the toy is at the top of the image so less thread shows.
3. Take several photos until you get one that looks real.
4. Post it online and see if people can guess how you got the photo to look real.

Remember: Don't try to fool the experts. Always post a follow-up shot revealing your trick and how you did it. Don't let anyone take you seriously.

7

Why Are Black Holes Black?

If light does not reflect off an object or if no light comes out of it, we see it as black. So if something is totally black, and it's out in black space, how do you see it at all? You can't. But you can tell it's there by watching things fall into it. Anything that gets close to a black hole is torn apart so violently by the black hole's super-high gravity, it heats up and shines brightly as it falls in, so we can see it through telescopes. That's what gives the black hole away.

What we do know is that a black hole is not really a hole; it's actually a solid object. If you could touch one, it would be the hardest, smoothest ball imaginable. But don't try it, because as soon as you get close, the black hole's gravity will suck you in and that will be the end of you. Some black holes have partners—ordinary stars that orbit around them. Because the stars are close to the black holes, the powerful gravity of the hole pulls material off the star and sucks it in. The black hole eats the star!

Have you ever noticed the shape that water takes as it goes down the drain of your sink? It swirls around and around, forming a hole in the center. You can see the same thing after flushing a toilet if you care to look. This shape, called a vortex, occurs over

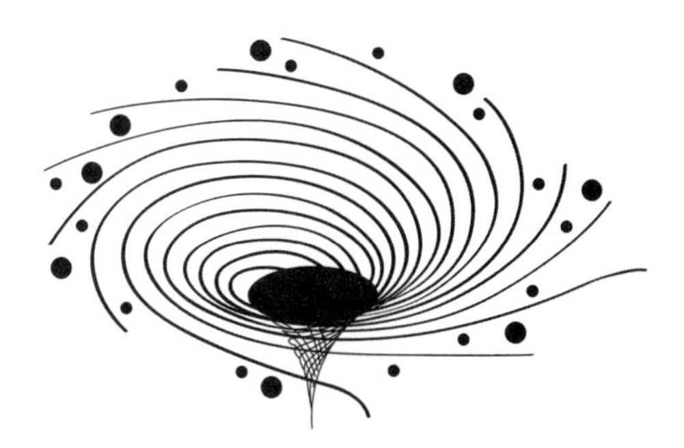

and over again in nature. Stuff gets caught and pulled in, and is never seen again. In April 2019, the first-ever image of a black hole was captured. It looked like an orange donut with a black spot in the center. That donut was the swirling disc of gas, and the black spot was the hole where it disappeared from our universe.

We see material constantly swirling around in space—gas and dust twirling around new stars; stars spinning around the center of the galaxy; galaxies circling around each other in big clusters. Close to a black hole, where gravity is extremely strong, everything that falls in spins around faster and faster until it disappears like water down a drain. Where all that material ends up, no one knows, which makes black holes one of the most mysterious objects in the universe.

And here's where the analogy to water in a drain falls short. We know that the water running down the drain doesn't really disappear; it just goes somewhere else. But not so in space. There's no plumbing or pipes up there, no tubes concealing what gets sucked in. When a black hole grabs something, it spins it around faster and faster until it reaches the speed of light, which (as far as we know) is the fastest anything in the universe can travel. Go faster than the speed of light and no one knows what happens because matter disappears from this universe as we know it.

Black holes don't come out of nowhere. They used to be stars, something like our sun, but much, much bigger and much heavier. All stars are balls of hot hydrogen gas that are trying to blow themselves apart. Stars want to explode outward, but they don't fly apart because their own gravity is pulling everything toward the star's center. As long as the outward push of explosive nuclear fusion and the inward pull of gravity have the same strength, the forces are completely balanced, and the star shines happily.

Stars come in many different sizes. Some are smaller than our sun, some are bigger . . . and a few are enormous—hundreds of times more massive than the sun. These are the stars that become black holes. Why? Because they have a lot more gravity.

Eventually, as all stars do, even the biggest stars stop burning on the inside. They run out of gas. When that happens, gravity takes over and the star is squeezed into a smaller and smaller ball. The smaller the ball becomes, the stronger its gravity gets, until something even stranger happens. The gravity becomes so strong that the light shining from the star can't escape, and the star turns black, disappearing from sight. But even though we can no longer see it, the black star is still there, and so is its super-powerful gravity. And it's that gravity that pulls in everything around it, including light itself. That's why black holes are black.

THE FUTURE IS NOW

What would happen if you fell into a black hole?

It would not be a pleasant experience, and it would be the last trip you ever made. The closer you get to it, the faster you fall. The point of no return in a black hole is called the event horizon. As you approach it, the powerful gravity would pull on your feet more than on your head, stretching you out into a long, thin string, like a piece of spaghetti. Then your body would be wound into a spiral and heated up until you were glowing white-hot.

Faster and faster you would spin until you reached the speed of light. The universe as you know it would disappear, and you'd be crushed to an incredibly tiny point in a place where time no longer exists. Some scientists think you'd end up in another universe, or perhaps pop out in another part of this universe. No one really knows.

If you want to look at a black hole, you can—well, kind of. If you're outside on a clear summer's night, look straight up over your head. You should see a big group of stars in the shape of a cross. This constellation is sometimes called the Northern Cross, but its true name is Cygnus the Swan. The star at the bottom of the cross is the swan's head. The star at the top is its tail. In between, the three stars making up the crosspiece form the wings. If you look carefully, almost exactly halfway from the head of the swan to the center of the wings, you'll find the spot where we discovered a black hole, Cygnus X-1, for the first time. Of course, you can't see the black hole because it's black (and really far away), but that is where it is located.

And if you look to the south, where the Milky Way widens as it touches the horizon, there's another group of stars shaped like a teapot. That's Sagittarius, and just to the right, or west, of the teapot, deep inside the Milky Way, is the galactic center—the home of the biggest black hole of them all. That super-giant hole is a million times more massive than the sun, and it's lying at the very heart of our Milky Way Galaxy. It's a good thing that one's really far away, because if it weren't, we wouldn't be here. Where would we be? Your guess is as good as mine.

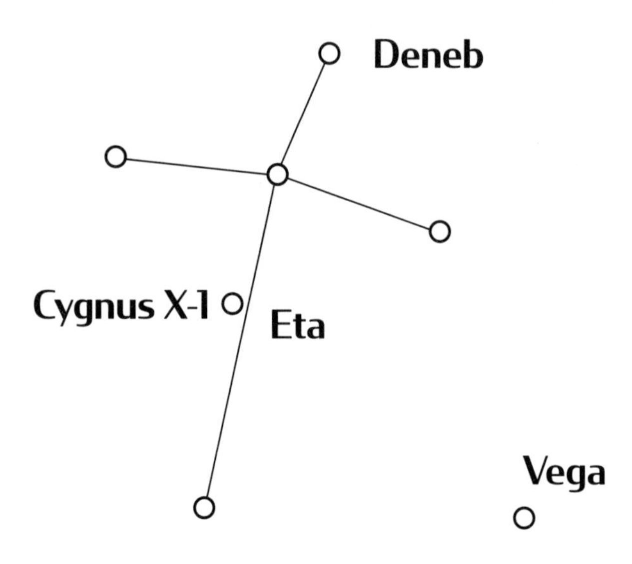

YOU TRY IT!

Bottled Black Hole

WHAT YOU NEED

1. Two clear plastic bottles with their caps (two-liter size works best)
2. A sharp knife or pair of scissors
3. Strong tape (black electrical tape is best)
4. Water

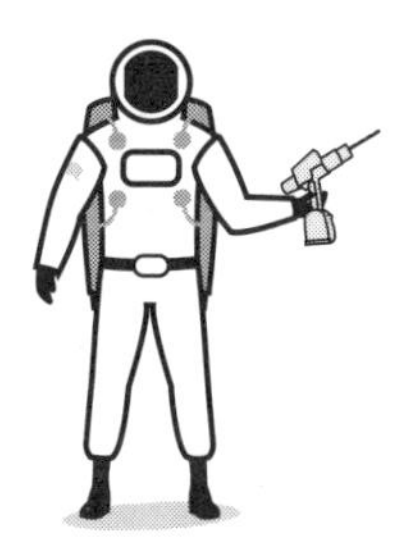

WHAT TO DO

1 Remove the labels from both bottles.
2. Using the knife or scissors, carefully cut a hole through each of the plastic caps, including the soft liners on the inside. The holes should be large enough so you can poke your little finger through them.
3. Hold the two caps so their tops are touching. Check to see that the holes are lined up so you can put your finger through both of them.
4. Wrap the tape tightly around the edges so the two caps are joined (don't cover the parts that screw into the bottle!). Make sure to wrap the tape around several times to attach the caps tightly.
5. Screw the joined caps tightly onto one of the bottles.
6. Fill the second bottle three-quarters full of water.

7. Screw the empty bottle tightly onto the one filled with water so the two bottles are joined together end to end.
8. Carefully pick up the bottle containing the water, making sure you don't twist or bend the joint between them. Then flip the bottles over so the one filled with water is now on top.
9. Swirl the bottles around in a circular motion to get the water spinning, and watch what happens as the water drains from the top bottle to the bottom one. You should see a long silver funnel appear in the water as it drains out. Once the funnel appears, stop swirling—the funnel will remain in place until all of the water is gone from the top bottle.
10. Once all the water has flowed into the bottom bottle, turn them over and repeat the action. You will see the same formation again. You can repeat this as often as you like.

You just made a model of a black hole! The funnel shape the water takes as it drains from one bottle to the other is a vortex. And remember: we can see that the water in your black hole model goes from one space into another, but we still don't know what happens to matter that falls into a black hole.

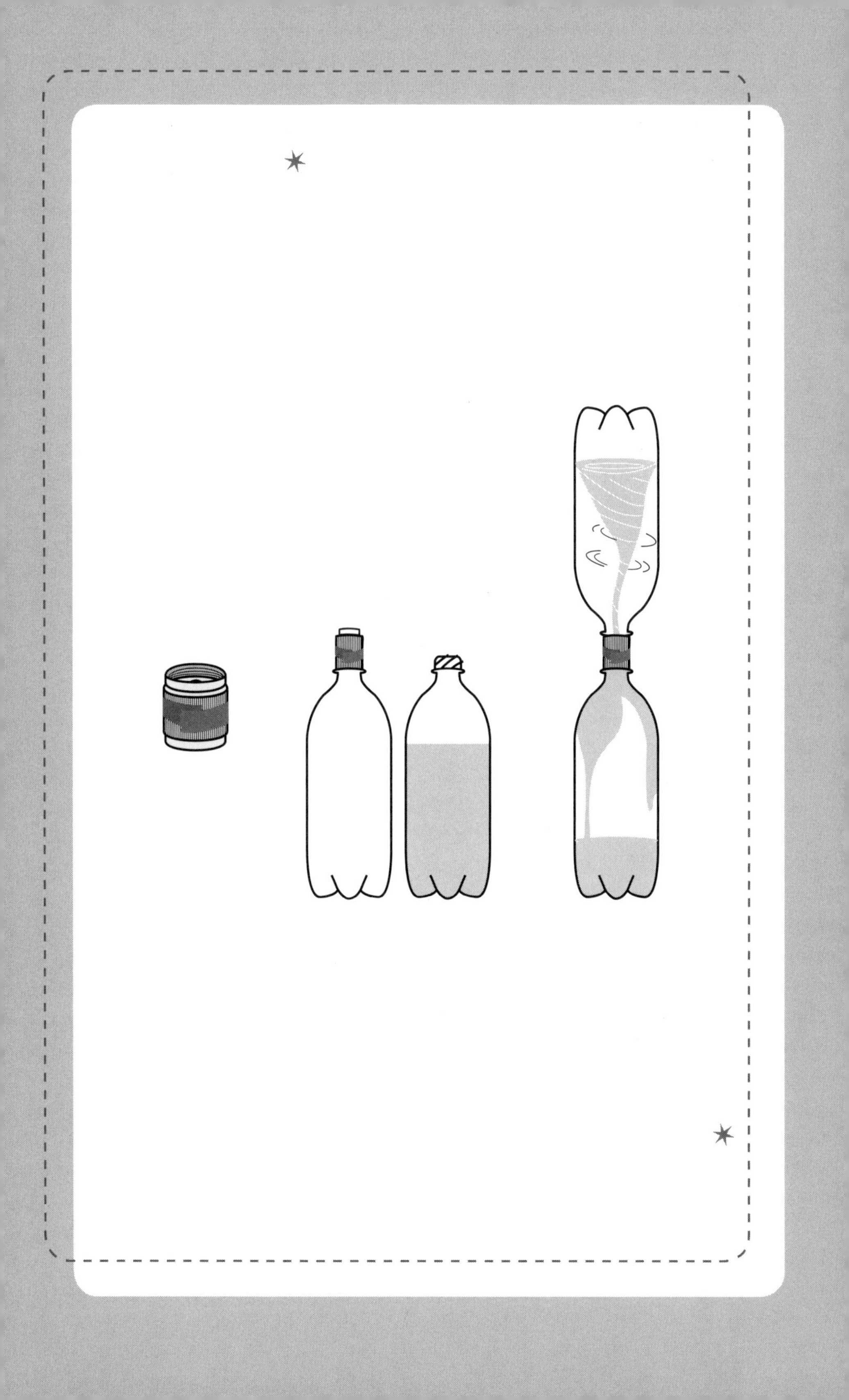

8

Why Do Some Planets Have Rings?

Whenever we draw pictures of planets, there is usually at least one with a ring around it. This is actually the way it is in space—some planets have rings; others don't. Alas, only the four biggest planets in our solar system—Jupiter, Saturn, Uranus, and Neptune—have rings. The four small planets—Mercury, Venus, Earth, and Mars—are ringless.

Why some planets have rings and others don't is a mystery. Perhaps rings are made when a moon comes too close to a planet and is torn apart by the planet's gravity. Perhaps two moons ran into each other and both were blown to bits, forming a ring of debris. Or maybe the rings are leftover stuff that never became a moon when the planet was first forming.

The planet with the most spectacular set of rings is Saturn, a planet that is dramatically different from ours. Earth is solid—it's made of rock. Saturn, though, is a big ball of hydrogen and helium, which are incredibly light gases. We use helium to fill party balloons. In fact,

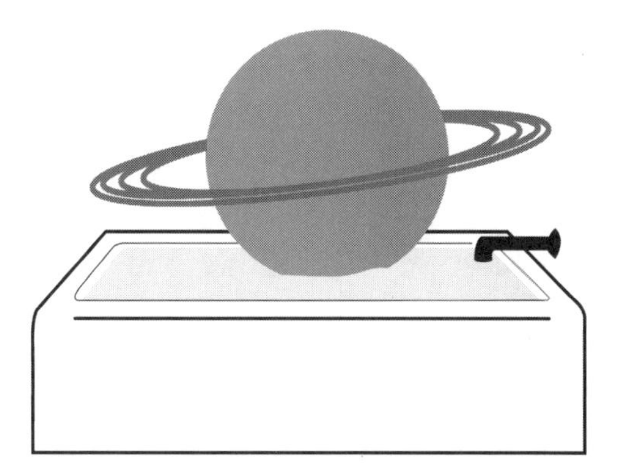

Saturn's the only planet in the solar system with a density less than that of water. That means if you had a bathtub big enough, Saturn would float. It would, of course, leave a ring.

The first person to see Saturn's rings was the amazing Galileo. He pointed a fairly small telescope at Saturn in the year 1610. And what he saw were what looked like handles sticking out either side of the planet. What was even stranger was that a couple months later, Galileo pointed his telescope at Saturn again, and the handles had disappeared. Galileo apparently said, "My glass has deceived me!"

What really happened is that Saturn had changed its position so that the flat rings had tilted up and Galileo was seeing them from the edge, like looking at a dinner plate from the side. Saturn's rings were so thin that they seemed to disappear when viewed with Galileo's small telescope. What could they possibly be made of?

As it turns out, the rings of Saturn are made of snowballs, ranging from about the size of your fist or your head to the size of a house. Billions of ice particles float around Saturn like the aftermath of a big snowball fight. They're spread out around the planet in a thin sheet that's hundreds of thousands of kilometers across but less than a kilometer thick.

The rings of Saturn are the most organized snowstorm in the solar system. Up close, you can see right through them. And if you were floating among all those snowballs, it would be easy to stick your head above them and see right across the whole system. What a sight that would be!

You might be wondering why the rings of Saturn circle the middle of the planet. You can find out why that is by using a tennis ball—any soft, unbreakable object, actually—and a long string.

Tie the ball to the end of the string and hold it straight out in front of you so the object is dangling just above the ground. The ball represents a ring particle, and the string is the planet's

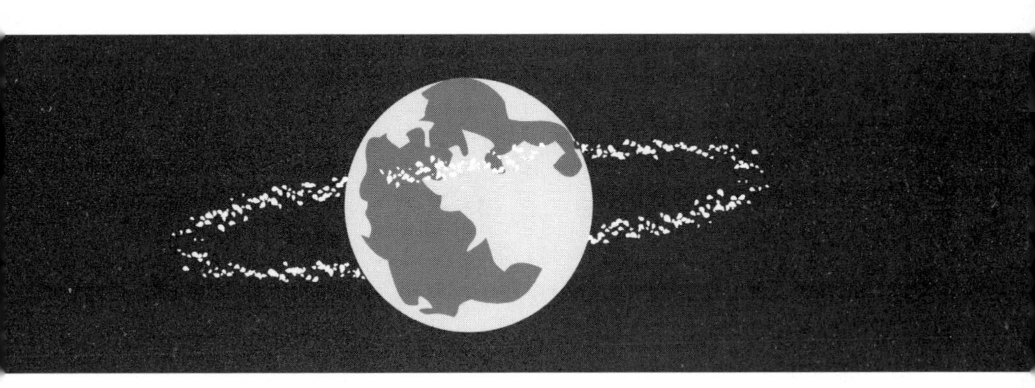

gravity. You're the planet! Think of your head as the North Pole and your feet as the South Pole.

All planets spin, but Saturn rotates very fast, twice as fast as Earth does, to be exact. So spin your body around on one spot while holding the end of the string out in front of you. Watch how the ball at the end of the string swings around with you. It's trying to fly away, but the string pulls it back, the same way that ring particles are trying to get away from a planet but gravity pulls them back. The faster you go, the higher the ball rises, but it won't go any higher than your arm holding the string, because that is where they can get the farthest away from the center of your body while spinning around. That's why the middle of a planet—its equator—is where rings are always found.

If you want to find out why rings don't rotate around the North or South Poles of a planet, hold the ball straight over your head. Now spin your body around the same way you did before. Maybe don't let the ball go, though—you'll probably get hit in the head!

When Saturn was becoming a planet, it formed from a huge cloud of gas and dust that was rotating like a flying disc. The stuff that was above and below the equator fell inward and became part of the planet itself. The stuff around the middle circled as fast as it could, but it wasn't drawn all the way into the planet. Instead, it remains circling around, caught in this gravitational tug-of-war.

The Earth doesn't have a ring today, but it did when it was younger. Billions of years ago, long before there was any life on our planet, scientists believe that a very large object, as big as the planet Mars, struck the Earth with a glancing blow.

The force of the impact almost ripped the Earth apart. A huge piece of our planet was sheared off and thrown into space, along with what was left of the object that hit it. For millions of years after this collision, an enormous ring of dirt and debris surrounded the Earth. Eventually, these ring particles clumped together into a ball.

That ball grew into what we now see as the moon. Hard to believe that the peaceful object that hangs so serenely in our night sky had such a violent beginning. Because our moon is far enough away from us, it isn't torn apart by Earth's gravity. But had the moon spiraled toward the Earth for some reason or formed much closer to the Earth than it is now, it would have been torn apart by our planet's tides, and we would have had a lovely ring system. In fact, it might even have been more spectacular than Saturn's—the moon is so large, it would supply a lot of material to spread out in a ring.

Amazingly, the rings that surround the four largest planets in our solar system are all different. Saturn's are made of ice and are very bright and wide. The rings around Uranus and Neptune are thin and dark, like pieces of charcoal, and they're arranged in long, thin lines with little moons running between them. The one around Jupiter is orange but barely visible. That ring is thin and made of dust that continually blows out of volcanoes on one of Jupiter's moons, forming a kind of smoke ring around the planet.

Still, the award for the weirdest rings in the solar system goes to Uranus. It has North and South Poles like the other planets, and it turns on its axis like the rest, but for some unknown reason, the planet is lying on its

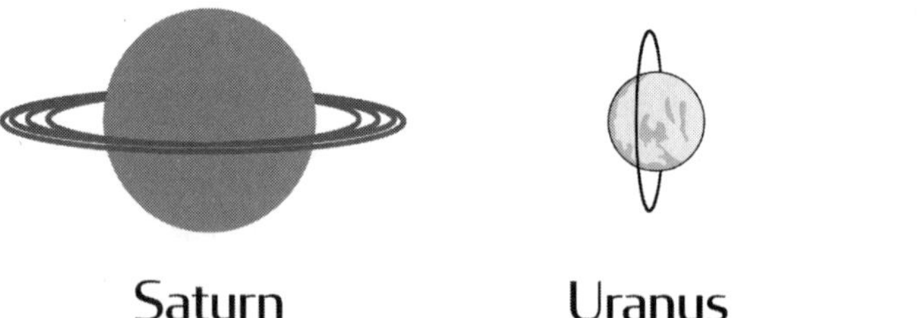

Neptune

side. It's as though the planet fell over and is rolling around the sun like a bowling ball. Uranus is the only planet that does this, and, oddly enough, its moons and rings go around the same way, circling the planet's equator.

So, the Earth once had a ring, which became our moon, and the rings around the other planets are all different from one another. That suggests rings form for different reasons. You can get a ring if a big object hits a planet, like the one that struck the Earth long ago. Rings might be leftover material from when the planet was forming, stuff that did not become a moon, or they can form when moons collide with one another. We know for sure that big planets have lots of moons. Jupiter has seventy-nine moons going around it, and Saturn has more than sixty. That creates a lot of opportunities for some of those moons to run into one another and get blown to bits, or for one of those moons to wander in too close to a planet, where the gravity of the planet will tear it apart and turn those pieces into a ring.

No one knows how long rings last. Perhaps they gather together into moons like the Earth's ring did, or maybe they eventually spiral into the planet and disappear. Recent data from a spacecraft called *Cassini*, which flew right between Saturn and its rings, found a "rain" of ring particles constantly falling into the planet. Scientists calculated that at that rate, the rings of Saturn will eventually disappear . . . in about 100 million years. That might mean that planets without rings are too small and don't have enough gravity to hold on to them, such as Mercury or even Venus, which has neither a ring nor a moon and has never been hit by something large. Or their rings might have disappeared over time.

Whatever the reasons, rings are among the most beautiful and mysterious objects in our solar system, so we're lucky we've got some in our backyard.

YOU TRY IT!

Put a Ring on It

You can build a scale model of Saturn and its rings.

WHAT YOU NEED

1. A large, clear glass bowl
2. Water
3. A spoon
4. A small sponge ball
5. Pepper

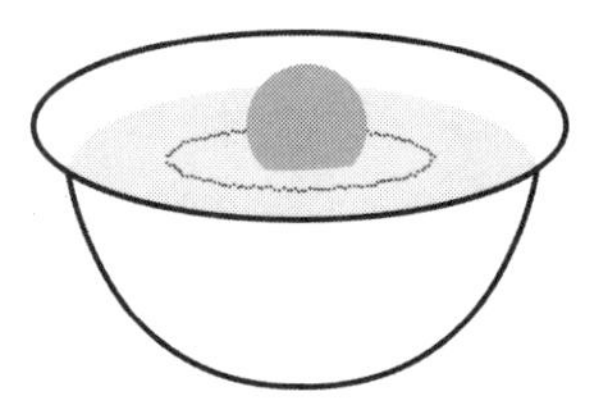

WHAT TO DO

1. Fill the bowl halfway with water.
2. Using the spoon, swirl the water around really fast until it's spinning in the bowl.
3. Drop the ball into the center of the spinning water.
4. Sprinkle the pepper around the ball until it covers the surface of the water.
5. You have just made a working model of Saturn's rings.

Can you see how all the particles swirl around the ball in the middle of the bowl? If you look closely, you can see that they move at different speeds: the ones close to the ball are moving faster than the ones farther away. That's exactly what happens around a planet. The ring looks like it's one solid piece, but all the little particles are moving on independent orbits.

Look through the side of the bowl. Notice how thin the rings are when you see them from the edge. They almost disappear when you look at them from the side. That's what fooled Galileo.

PART 2

Home, Sweet Home

Our Planet Earth

9

What Is Our Cosmic Address?

We live on a planet . . .

That is one of seven other planets . . .

That circle our sun . . .

Which is only one star among billions of others . . .

That form our galaxy, called the Milky Way . . .

Which is only one galaxy among billions of others . . .

That are scattered across the expanding universe. . . .

Another way of putting it: we live on a ball.

It took a lot of science to figure out that the Earth is round, because to all of our senses, it looks flat. The ground is always beneath our feet, and the sky is always above our heads. Continents are basically just big islands, so if you walk in any direction and keep going for long enough, you'll come

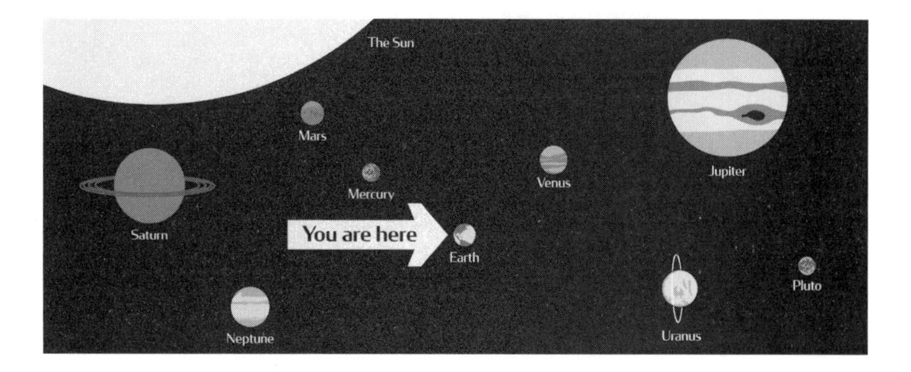

to an ocean. And when you look out across the ocean, the horizon where the water meets the sky appears to be very flat. In fact, it looks like you could fall off the edge if you went too far.

Thousands of years ago, ancient sky watchers believed that the Earth was an island in a huge ocean, and that the sky was an enormous dome that rotated over our heads every day. Half of the dome was blue, with the sun stuck on it; the other half was black with the moon and stars on it.

This model of the Earth makes sense only if you think about how we see the world with our five senses. But we are limited as human beings. We can't see far ahead of us, and we're very small compared to the Earth itself. And if you are just a small being on a very large ball, the surface of the ball will look flat.

Thankfully, we learned to sail. As a result, we realized we could travel all the way around the world in one direction and still return back to the point we started at, proving the Earth is round. Later, we flew around it in airplanes, and eventually, we left Earth's atmosphere altogether and saw our beautiful little planet from space. And yes, it really is a ball!

Even though we have all of this scientific evidence to show that the Earth is a ball, there are still some people who believe the Earth is flat. Members of the Flat Earth Society believe that our planet is shaped like a pancake, with the oceans running around the outside. It's similar to the ancient view of the Earth, which is based only on what we can see with our own two eyes. But if that's all you use to see the world, you are missing a lot. Science is another set of eyes that lets us see beyond the flat horizon, out into space, and back at our beautiful blue ball floating in the blackness of the universe.

To get a sense of our place in the galaxy, it helps to imagine a scale model. In fact, if you wanted to build the model yourself, just about everything you need can be found in your kitchen.

Find a grapefruit, some poppy seeds, some sprinkles used to decorate cakes, and some round candies, such as Smarties or M&M's. Once you have everything, head to a park with a baseball diamond, along with some of your friends.

At home plate, compare the size of the grapefruit to one of the candy sprinkles. That's about the size difference between the sun and the Earth.

Place the grapefruit on home plate to represent the sun at the center of the solar system. Walk toward the pitcher's mound and stop halfway. Place one poppy seed on the ground there. That's the planet Mercury. You might want to place it on a piece of paper so you can see it.

Continue walking and place one candy sprinkle on the pitcher's mound. That's Venus. Drop another one at second base to be the Earth, and another about twenty paces out to be Mars. Now walk all the way out to the fence at the edge of the outfield and place one of the round candies there to represent Jupiter.

Look back at how far away the grapefruit is and how tiny all the planets are. And you are *still* not out of the solar system!

Saturn will be another round candy about a block away from the park; Uranus and Neptune will each be another block after that. You should now be about a kilometer away from the grapefruit sun. Congratulations—you've created the ballpark of the solar system!

The next-closest star to our sun is Alpha Centauri. If Alpha Centauri were a grapefruit, it would be four thousand kilometers away from home base in our model—that's the distance between Toronto and Vancouver!

Of course, we're not sitting still in space. The Earth is always moving. In fact, the Earth moves in seven different ways:

1. The Earth spins 360 degrees once every day (twenty-four hours). That means the Earth is carrying you around like you're riding a carousel. You're traveling very fast around the center of the Earth, no matter where you live. Most cities in Canada are moving at about eight hundred kilometers per hour—the speed of a jet airliner. A person living near the equator is moving 1,600 kilometers per hour—faster than a supersonic jet fighter!

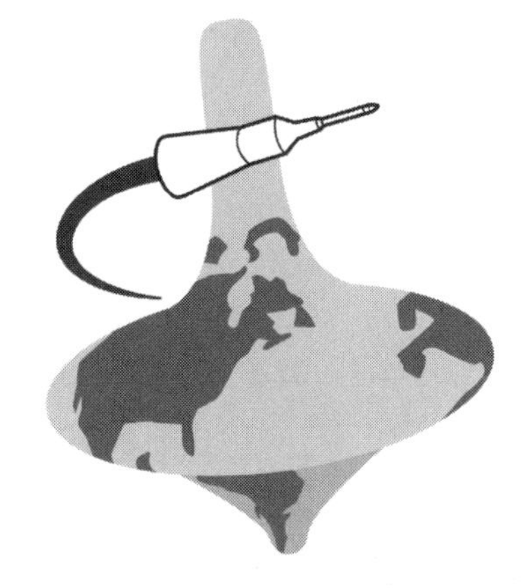

2. The Earth wobbles like a top, with the North Pole tracing a circle in the sky once every twenty-six thousand years.
3. Our planet nods up and down like a big head slowly saying yes. It does this every month because the moon goes around the Earth at an angle and the gravity of the moon pulls the Earth up and down.
4. The Earth circles around the sun every year, moving at a speed of one hundred thousand kilometers per hour. That's thirty kilometers every second, which is faster than most rockets.
5. Our path around the sun is not a smooth circle because the moon pulls us from side to side as it circles around us. Imagine swinging a heavy weight on a string around your head while walking down the street. The weight will make you stagger from side to side. The Earth follows a wavy path around the sun, with twelve waves in it, representing the twelve months of the year.
6. The same way that the Earth revolves around the sun, the sun itself is traveling around the center of the Milky Way Galaxy. The sun's speed is just over a million kilometers per hour, yet it still takes 250 million years for it to circle the galaxy once, carrying us and the seven other planets in our system with it. The last time we were on this side of the galaxy, dinosaurs were just emerging.
7. Our Milky Way Galaxy is one of a small group of about fifty other galaxies that all move around each other, and our local group is one of billions of other such galactic groups scattered across the vastness of the universe, which is itself expanding, getting bigger and bigger every moment. It's enough to make you dizzy!

It has taken thousands of years of sky watching, telescopes, astronomy, and spaceflight to figure out our cosmic address. But knowing our place in the universe will help us as we explore the cosmos. Just don't expect the mail from our neighbors to arrive any time soon.

YOU TRY IT!

A Guide to Earth's Orbit, for Aliens (and Earthlings)

There are two times in a day when you can watch the Earth turn: at sunrise and sunset. While it appears that the sun is sinking into the horizon at sunset and rising from it at dawn, the fact is that the Earth is turning, not the sun, and it's carrying you with it.

WHAT YOU NEED

1. A lamp (a desk lamp, or one with a shade that shines light in one direction, is best)
2. A swivel chair, or a stool
3. A dark room

WHAT TO DO

1. Place the lamp in the middle of the room with the chair a short distance from it. Aim the lamp at the chair. Then sit in the chair and face the light.
2. Think of your head as the Earth and the lamp as the sun. When you face the lamp, it's daytime.

3. Slowly turn your body and your head in a counterclockwise direction and notice how the "sun" seems to be moving to the right. This is the direction the sun moves across the sky during the day when you're facing south. It rises in the east and moves to the right across the sky until it sets in the west.
4. Keep turning around until the sun is on the edge of your vision on your right. That is sunset.
5. Turn more and you'll not see the sun at all. Now you are in night, or on the dark side of the Earth. Continue all the way around and the sun will reappear on your left side, which is sunrise. Welcome to a new day.

If you want to give the experiment a twist, try repeating it from the opposite side of the lamp. You should be looking back at where you were sitting before. Notice how now, when it's "night" and your back is toward the light, you're facing a different direction than you were before? That is why the summer constellations are different from those in winter. Stars completely surround the Earth, just as the walls of your room are all around the lamp. We face opposite directions in the sky (or different "walls") in summer compared to winter, which is why we see different stars in each season.

10

How Big Is the Earth?

Compared to a human being, the Earth is enormous. If twenty million people (almost half the population of Canada) held hands with their arms stretched out, they would reach around it once. If you wanted to walk all the way around the planet, you would have to cover 40,074 kilometers. It would take two years of nonstop walking, twenty-four hours a day, to make the trip. That's quite the hike!

The Earth looks flat to us because we're so small and stuck on its surface. If you hold a big ball, like a basketball or volleyball, right up to your eye and look at the edge, the ball doesn't look as round. The only people who get to fully appreciate the roundness of the Earth in person are astronauts who fly high above it in space.

Today, we take seeing the Earth from space for granted. But it wasn't always this way. It took many scientific minds and centuries of technological

innovation to get us to where we are. One of those early science pioneers was a man named Eratosthenes, who worked in the great library of Alexandria in Egypt more than two thousand years ago. He made the first accurate measurement of the Earth, not only proving that the Earth was round but also showing how big it was. The most amazing part? He did it all with just a stick.

One day in June, Eratosthenes was in his hometown, Syene, along the Nile River in Egypt. As he passed by a water well, he did what we all do when we pass a well: he looked down to see the water at the bottom. Not only did he see the water, he saw a reflection of the sun, which meant that at that moment, the sun was straight overhead.

Now, I'm not sure that walking by that well was an accident, because he did this at exactly noon on June 21, the summer solstice, the one day of the year when the sun is at its highest point in the sky. If the Earth were flat, then at noon on June 21, Eratosthenes would have been able to see the sun at the bottom of every water well throughout the world, because sunlight falls straight down from the sky. But if the Earth were round, he'd only see the sun in one well at a time and only when the sun was directly above each one.

So Eratosthenes reasoned that if he could measure the angle of the sun at another well at the same time on the same date, he could tell how far around the curve of the planet he was . . . and, by a simple calculation, figure out the size of the Earth.

Eratosthenes waited a year to do the measurement. Just before noon on June 21 the following year, he stepped outside the library in Alexandria, several hundred kilometers north of his hometown. There was no water well to see the sun in, so to find out whether the sun was straight overhead, he used a stick. Not just any stick, but a very special one called a gnomon—it's like a sundial that measures the angle of the sun in the sky.

Right before noon, he put the gnomon on the ground and waited for the sun to pass overhead. If the shadow of the stick disappeared entirely at noon, it meant that the Earth was flat—end of experiment. But of course, the shadow didn't disappear. As the sun passed overhead, the shadow shrank to a little sliver, but it never completely went away. When Eratosthenes

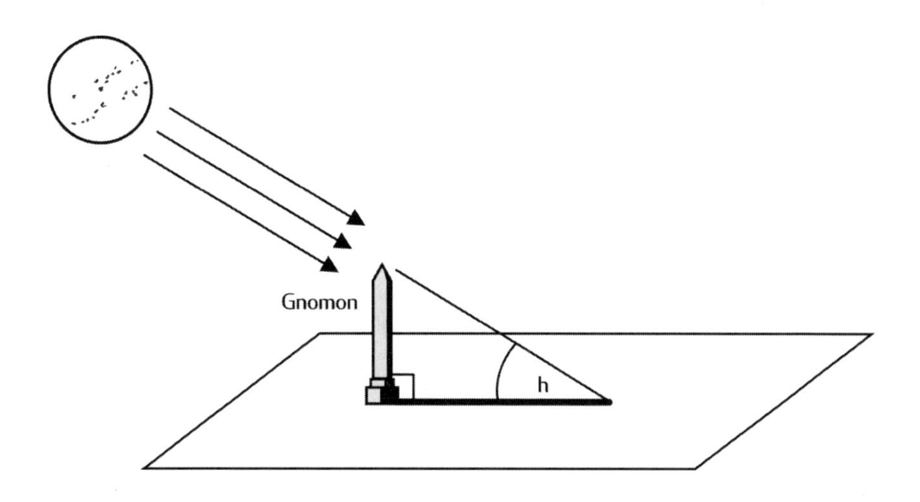

measured the angle of the shadow, it was seven degrees. It wasn't much, but it proved that the Earth was curved.

What Eratosthenes had measured was a thin slice of a circle, like a piece of pie. Seven degrees is about one-fiftieth of a circle (a circle has 360 degrees). That meant that the distance between the towns of Syene and Alexandria was about one-fiftieth of the distance around the Earth. He knew how far apart the two towns were, so he just multiplied that by fifty.

In ancient Greece, length was measured in units called stadia—as long as a stadium, or about two hundred meters. (The stadium was where people gathered to watch races, just as we still do today.) Eratosthenes figured that the Earth had a circumference of two hundred fifty thousand stadia, which translates to roughly fifty thousand kilometers. The actual distance around the Earth is forty thousand kilometers, so he was pretty close!

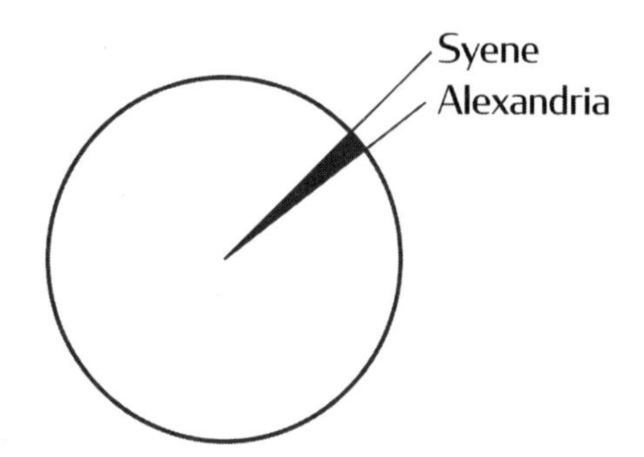

Eratosthenes was the first person we know of to get a sense of how large our planet really is. Imagine what went through his mind more than two thousand years ago, realizing that the world

was round and that it was far bigger than what anyone knew at the time. The ancient Greeks didn't know that North America, South America, Australia, and Antarctica existed, because ships at that time could not cross the oceans.

Later, other civilizations built upon the discoveries of the Greeks. In 1000 AD, a Muslim mathematician named Al-Biruni found a new way to measure the curvature of the Earth by sighting along the ground to the top of a mountain. Using trigonometry, the branch of mathematics involving triangles, Al-Biruni was able to calculate the distance to the center of the Earth and, from that, figure out the planet's size.

Despite how far we've come since Eratosthenes, we need to keep things in perspective. Although the Earth seems big to us, it is actually small as planets go. Just look at the Earth compared to Jupiter, the largest planet in our solar system. A thousand Earths would fit inside Jupiter. And if we could place the Earth at one side of Saturn's rings, the moon would be on the other side.

Most planets that have been found orbiting other stars in the universe are big, like Jupiter and Saturn. Our planet may be small, but it is the only one we have!

YOU TRY IT!

Eratosthenes's Experiment

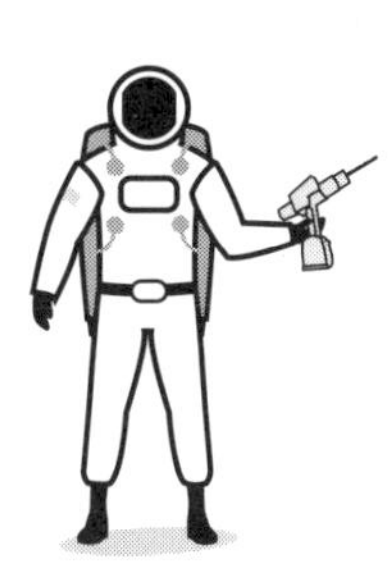

WHAT YOU NEED

1. Half of an 8x10 picture frame that forms an L shape
2. A friend in another location with the same thing, same size
3. Measuring tape or a ruler
4. Phone

WHAT TO DO

1. On a day that is sunny at both your location and your friend's, call your friend.
2. Both of you should go outside and place the picture frame on something flat, such as a windowsill or a patio. It is very important that the surface is level so one arm of the frame points straight up.
3. Turn the bottom part of the L until the shadow of the upright piece falls along it.
4. At the same moment, both of you measure the length of the shadow using the ruler.
5. Compare your measurements.

There will be a difference in the length of the two shadows because of the curve of the Earth. The farther away you are from each other, the better.

11

Why Are Planets Round?

Look at anything in space and what do you see? All the planets are shaped like balls. So are the sun and all the other stars in the universe. Planets circle the sun in round orbits, and the sun follows a circular path around the Milky Way Galaxy. Everything seems to be round. And there's one force out there making it work that way: gravity.

Everyone on Earth feels the pull of gravity. It doesn't matter where you are—you feel a pull toward the ground, and you think you're at the "top" of the planet. In fact, there is no single up or down; each one of us has our own personal sense of where down is, and all of us are correct! Down is simply toward the center of the ball.

Look at a globe of the Earth and find your home country—say, Canada. If you turn the Earth over, you'll find Australia on the other side. That means at this very moment there are people under your feet, on the other side of the world, who are calling "down" what you call "up."

Everything on the planet is pulled toward the middle of the Earth. The only shape that allows every part to be as close to the center as possible is a ball. Take a piece of Silly Putty, or if it is winter, make a snowball, and try to make it as small and compact as possible. Squeeze everything toward the center as best you can. What do you end up with? A ball.

Gravity works the same way, always pulling toward the center of an object or a group of objects, gathering everything into the smallest possible shape. That's why large objects in space, no matter what they're made of, are round.

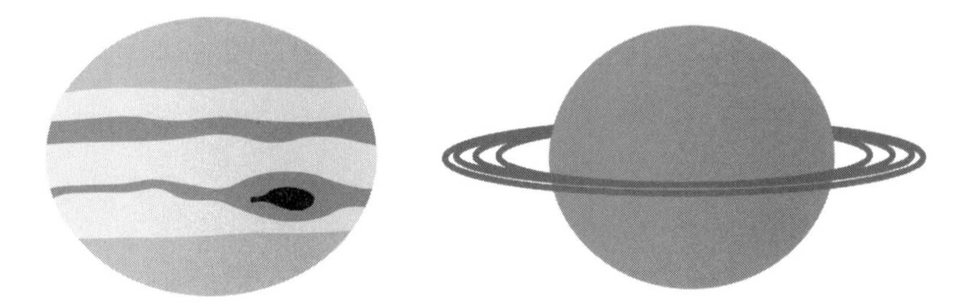

The bigger an object is, the more gravity it possesses. By the way, we don't know where gravity comes from. We just know that the more mass you have, the more gravity you get. As gravity continues to pull objects in toward the center, we get a larger and larger object . . . and the larger it grows, the more gravity it has, which pulls even more stuff to it.

Our whole solar system works this way—small objects gravitationally circling big ones. The sun is the biggest of all, thousands of times bigger than all the planets put together, so it has the most gravity. All the planets—which are little by comparison—are caught in the endless circle of the sun's gravity. At the same time, moons are smaller than planets, so they loop around the planets as the planets circle around the sun. Everything is swinging around something in a giant circular swing dance in space.

All of those forces have an effect on the way the stars and planets move around the galaxy, which is also spinning around its center. But, although all planets are shaped like balls, they're not always perfectly round. Sometimes they can have a little bulge in the middle.

The biggest planets in our solar system have the biggest bulges. Jupiter, Saturn, Uranus, and Neptune are called gas giant planets because they're giant compared to the Earth and they're made mostly of gas and liquids.

Imagine a liquid ball floating in space. Gravity is pulling it together into a sphere, but big planets rotate very quickly, twice as fast as the Earth. And when a ball of gas or liquid spins around, the equator is moving faster than the poles, the same way that the tire of a bicycle moves faster than the hub.

That causes the planet to bulge out at the center because its material is being flung outward by the spin.

The Earth also bulges at the equator—when you stand on the equator, you are forty-two kilometers farther away from the center of the Earth than you would be if you stood at the North Pole. But that's nothing compared to Jupiter, which is made of gas. Measured across its middle, Jupiter is 143,884 kilometers. But when you measure the planet from north to south, it's 133,709 kilometers. That means that Jupiter's equator sticks out an extra 10,175 kilometers farther away from its center. If it didn't spin at all, Jupiter would be perfectly round.

Really small masses—say, less than one hundred kilometers across—come in crazy shapes. Space masses that small don't possess enough gravity to pull everything toward the center and form a ball. Phobos, a moon of Mars that is only twenty-two kilometers across, looks like a potato. And an asteroid named Eros, which is about the same size, is shaped like a peanut. There's even a comet that looks like a dog bone! These odd-shaped little moons and asteroids could be pieces that were chipped off larger bodies, or stuff that just never formed into a big moon or planet.

Recently, a robot spacecraft called *Rosetta* caught up with a comet that looked like a rubber duck. *Philae*, a small lander that was dropped off by *Rosetta*, was sent to the surface of the duck comet to try to take a sample. *Philae* had a harpoon in its belly and claws on its feet that were supposed to dig into the comet so the lander would not bounce off in the low gravity.

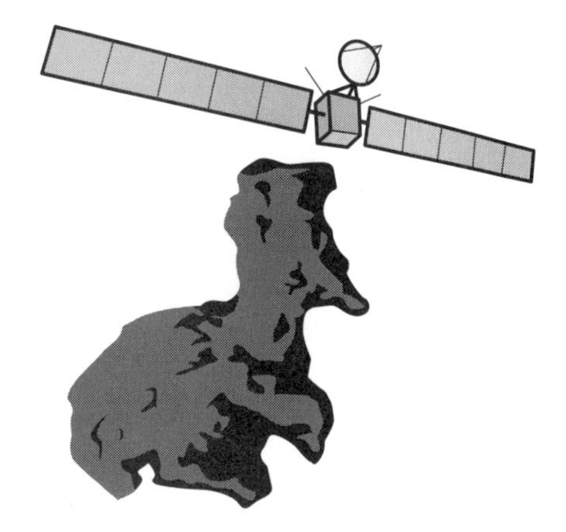

But there was a bit of a mishap. The harpoon didn't fire, and the claws didn't hang on, so the lander hit the ground and bounced back up. With almost no gravity at all on the comet, *Philae* drifted back up a kilometer high above the comet, then slowly floated back down and ricocheted up a second time. Each bounce took about a minute to happen. Finally, *Philae* ended up tilted at a crazy angle against a cliff.

If astronauts ever visit a comet or asteroid, hopefully they won't bounce like *Philae*, but they will be able to take giant, slow-motion leaps and fly just like superheroes. Wouldn't that be fun?

For now, we know how gravity works—it pulls you down, tries to make you round, and keeps you stuck here on the ground. But we still don't know what it is beyond a mysterious, invisible force. If we understood it better, maybe we'd be able to turn it off, or turn the volume up and down like on a TV. Imagine that! Maybe one of *you* will become the scientist that figures out gravity and makes it work by remote control!

YOU TRY IT!

Garbage Gravity

You can see how gravity makes objects round by using a garbage can gravity well.

WHAT YOU NEED

1. A large, round garbage can
2. String
3. A thin tablecloth or sheet
4. A bag of marbles

WHAT TO DO

1. Drape the cloth over the top of the garbage can and tie it tight around the rim with a piece of string so it looks like a drum. Make sure the surface is smooth and the fabric is taut. The surface of the cloth represents space.
2. Roll two or three marbles across the drum and notice how they move in fairly straight lines. You will also see how they end up stopping in the center and their weight causes the fabric of the drum to bend downward, forming a little well. Albert Einstein described gravity as the curvature of space—every object causes space to bend inward, causing other objects to follow that curve until the two come together.

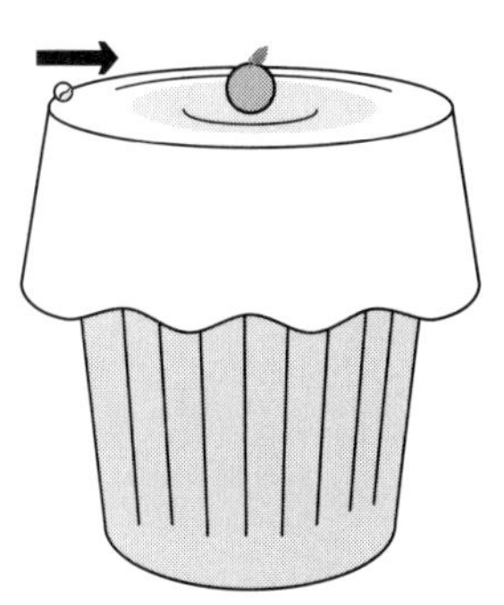

3. Repeat the above step with the rest of the marbles one at a time until they all cluster in the center.
4. When you have created a deep well, roll one marble around the cluster so it goes into orbit. You have just made a model of a planet going around a star!

The more marbles you add, the more the fabric bends. Look down from above. Can you see how all of the marbles gather together and make a round shape? That's how gravity makes stars and planets round: by pulling everything as close to the center as possible.

12

How Do We Know the Earth Moves?

The ground beneath your feet doesn't feel like it's moving. Today, we know that the Earth is spinning like a top, whizzing around the sun, and circling the galaxy at dizzying speeds. But hundreds of years ago, people thought that the Earth was at the center of the universe because everything in space seemed to be moving except us!

Galileo was an astronomer who lived more than four hundred years ago and was the first to use a telescope to look at the moon and the planets. He was also the first to prove that the Earth really does move around the sun. His biggest clue came when he saw Jupiter through his telescope and

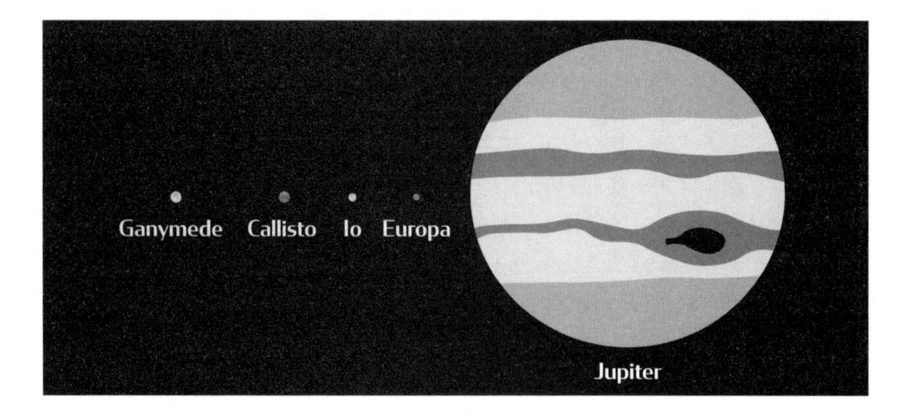

picked out four tiny dots on either side of the planet. As he observed Jupiter night after night, the dots changed their positions. He measured their motions and calculated that the dots were moons orbiting the big planet. Those four moons are now called the Galilean satellites.

Galileo reasoned that if small moons orbit a big planet, then it makes sense that small planets would go around our big sun. He was right, but his discovery got him into trouble for the rest of his life.

Back in Galileo's time, most astronomers—and the Pope, who was the powerful head of the church—believed that the Earth didn't move at all. They thought that the Earth was at the center of the solar system, with the sun, moon, and all the planets going around it. It's a nice idea, thinking we're so important, but it's wrong.

We now know that the sun is at the center of our system, and we go around it. Although all of the planets in our solar system orbit the sun, those orbits aren't completely even, and we occasionally pass the other planets, like one car overtaking another on the highway. Take Mars, for example. It's farther away from the sun than we are, so it takes longer to complete its orbit. Every now and then, the Earth passes Mars like we are on the inside lane of a racetrack, and when that happens, the red planet can be seen closer in our night sky, and sometimes it seems to move backward as we pass it. Jupiter and Saturn do the same thing, although they're farther away and don't seem to back up as much.

That's why early models that tried to duplicate the motion of the planets with the Earth at the center got so complicated. Ancient scientists were trying to represent this strange motion with a model that had the Earth standing still. Oops.

Galileo was different. He believed the Earth *did* move. He wasn't the first to come up with the idea, but he was the first to prove it using a good scientific method. He knew the key to understanding how the Earth moved was gravity, but he wondered: Would gravity work the same way on a large mass like the Earth as it does on a small mass?

To test his idea, he went to the Leaning Tower of Pisa.

The story goes that Galileo threw a bunch of objects off the tower to see if they would fall at the same speed. He was trying to better understand

the laws of gravity. He figured that if objects of different sizes fell at the same speed, it would mean that gravity worked the same way on all objects on Earth, no matter what size they were. And if that was the case, Galileo figured that gravity would also work the same on the planets in the solar system.

Galileo's experiments proved his hunch. With that theory of gravity in mind, Galileo found that the motions of the planets around the sun all made sense. The gravity of the sun held on to the planets as they orbited around it. In other words, the Earth did not sit still; it moved.

Galileo wrote the results of his experiments in a book that became an instant bestseller. Scientists liked what they read. They started pointing their own telescopes into the sky and agreeing with his idea that the sun was at the center of the solar system. Galileo, in other words, started modern astronomy.

There was just one little problem with Galileo's new idea that the Earth moved. The Pope at the time, Urban VIII, also happened to be an astronomer . . . well, sort of. He believed the official church position that the Earth was unmovable, and therefore all the stars and planets had to go around us. In the Pope's eyes, we were the center of the solar system, not the sun. And the Pope did not like being told he was wrong.

Galileo was put on trial for his discoveries. The Church threatened to throw him in jail or even torture him if he didn't take back what he said. Fortunately, Galileo was famous enough that he escaped torture and imprisonment, because the Church and state feared the people would protest.

Galileo was forced to take back his ideas and appear contrite. Still, he never actually said he was wrong. He merely said his observations could be interpreted differently. That satisfied the Church, but deep down, Galileo knew he was right.

His punishment was to be confined to his home, a fairly nice place in the hills outside of Florence, Italy. In fact, you can still visit his house today. The only problem was that he wasn't allowed to leave. Imagine being told you can't leave your house for the rest of your life. It was way better than jail or torture, but it was still limiting. Beyond confinement, Galileo also wasn't allowed to do experiments or use a telescope. Of course, that didn't stop him from working on his ideas and quietly passing them on.

Galileo had young students visit him at his home, and he gave all of his knowledge to the newer generation, hoping that they would one day carry on his research. And that's proof that he was truly a great scientist. He, like so many other gifted and learned men and women of science, passed his knowledge down to others.

There's an important lesson that Galileo taught us. He believed he was right while others did not. He stood up for his beliefs and did everything he could to spread his ideas to others. We remember him for that.

So if you know you're right about something, stand up for it. Don't let people who are wrong try to change your mind. You might be criticized for your thinking, but in the end you'll be remembered. You might even change the course of history!

That's certainly what happened with Galileo. Those little moons that he saw with his telescope in 1610 have turned out to be four of the strangest places in space. All of them are roughly the size of our moon, but they're very different from what we typically consider to be a moon. And one of the robot spacecrafts that went to Jupiter to study these moons was called *Galileo*!

First of all, the moons come in different colors. The one closest to Jupiter is orange. The next one out is white. And the other two are shades of gray.

Each moon has been given an interesting name—Io, Europa, Ganymede, and Callisto. Much better than the name we gave our moon . . . "*the* moon."

Io is the strangest of the four moons. It's covered in orange and yellow crystals made of sulfur that blows out of volcanoes and falls like multicolored snow on the ground. The whole moon seems to be turning itself inside

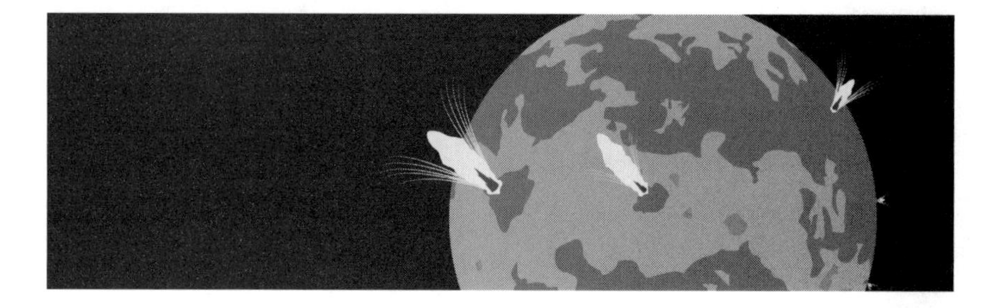

out as hot material from the inside spews out of holes in its surface, forming umbrella-shaped clouds that rain particles down onto the ground.

In some places on Io, the ground is so hot, it has melted into black lakes of liquid goo with yellow islands in them. It looks like something out of science fiction, but it's all real. It's a colorful, violent place.

Europa is the opposite of Io: it's a frozen world. The whole surface of this moon is one huge sheet of ice. If you could hold Europa in your hand, it would feel like a perfectly round ice ball. There are no big mountains or valleys on this moon, just cracks in the ice. Some of the pieces of ice look like they've moved around; some of the cracks look fresh and new.

Ganymede and Callisto look more like our moon, except they're made of both rocks and ice mixed together. Ganymede has huge circular areas that look like impacts from large objects, and patterns in the ice suggest it is floating on an ocean of liquid water. Callisto has more craters on it than any other moon, which suggests it hasn't changed much in a very long time.

You can see the moons of Jupiter with a good pair of binoculars or a small telescope.

If you can't get one, try attending a star party hosted by an astronomical society. The parties are usually free, and you can look through their high-powered telescopes to see the weird and wonderful world beyond our atmosphere. And when you do, give a shout-out to Galileo for making it all happen!

YOU TRY IT!

Look Out Below

You can do the same experiment Galileo did four hundred years ago. And you don't need to climb a big tower to do it.

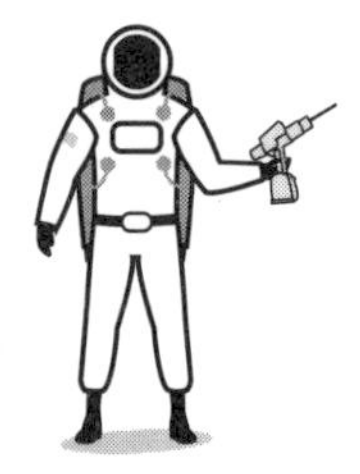

WHAT YOU NEED

1. A large, heavy book
2. A flat sheet of paper

WHAT TO DO

1. Hold the book in one hand and the paper in the other, with both of them lying flat.
2. Drop both objects at the same time and watch to see which one hits the ground first.
3. The book should win the race to the ground.

Galileo said that all objects should fall at the same speed, so does that mean Galileo was wrong about gravity? Not really. When you drop the piece of paper, it falls slowly because it is very light and it is slowed down by the air. If there were no air, though, both the paper and the book would fall at the same speed. Don't believe me? Let's try the experiment one more time.

Repeat the experiment, but this time, lay the paper flat on top of the book and drop both of them together. What happens? The paper falls at the same speed as the book because the book pushes the air out of the way.

Try the experiment one more time, and this time crush the paper into a ball as small as possible, then drop it beside the book.

You will see that both objects do fall to the floor at the same speed because the balled-up paper does not catch as much air. So when you take air out of the equation, you can see that Galileo was right. Gravity acts on all objects the same way, whether it is a book and a piece of paper, the moons of Jupiter, or the planets going around the sun.

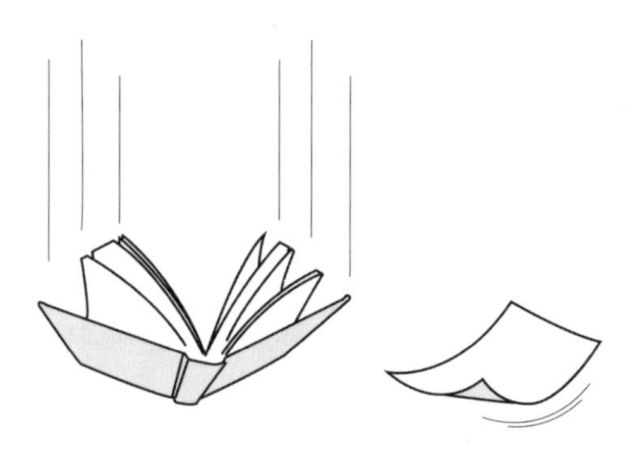

13

What Do Satellites Do?

A satellite is any object that orbits a planet. The Earth used to have only one satellite: the moon. But since the invention of rockets, we've sent more than four thousand satellites up into space to orbit our planet, and hundreds more are launched each year.

The thousands of satellites orbiting Earth every day are watching the weather, relaying television programs around the world, monitoring the environment . . . Even your phone talks to satellites. The GPS in our phones relies on a whole fleet of satellites that send out signals from space that your phone captures to find your location anywhere on the planet. When you hit the map function, your whereabouts are accurate because your phone is communicating with satellites orbiting over your head.

Canada was the fourth country in the world (after Russia in 1957 and the United States a year later) to launch a satellite. Our first was *Alouette 1*,

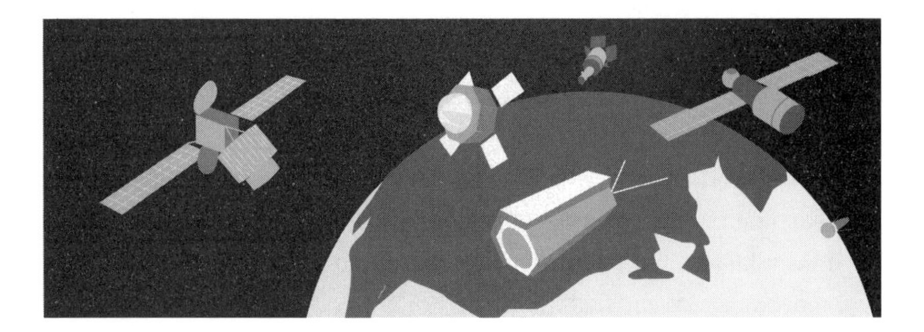

which studied the upper atmosphere of the Earth, far above where airplanes can fly. *Alouette 1* is no longer working, but it's still up there.

There are so many satellites circling the Earth that you can spot one yourself. Start looking just after it gets dark—that's when the sun is still shining on the satellites, making them look like moving stars. You have to be patient, though. Don't be fooled by airplanes. If the moving dot has a blinking light on it, that's a plane. Look for something that appears to be a moving star—it probably won't be very bright—drifting slowly and smoothly across the sky. If you spot a really bright one, that could be the International Space Station.

Satellites orbit the Earth in a variety of ways, depending on their function. Some fly low, others go high; many go around the equator of the Earth while others loop over the North and South Poles. They may carry cameras to look down on our planet or telescopes to peer out to the edge of the universe.

When a satellite carries people, as in the case of the International Space Station, it orbits around the middle of the Earth at an angle across the equator, arcing high over North America and Russia, then down toward South America and Australia. This orbit is called low Earth orbit (LEO). It's the easiest orbit to reach because it's not very high—roughly four hundred kilometers up, which is barely above the Earth's atmosphere—and it offers the best views for astronauts as they pass over most large cities in the world.

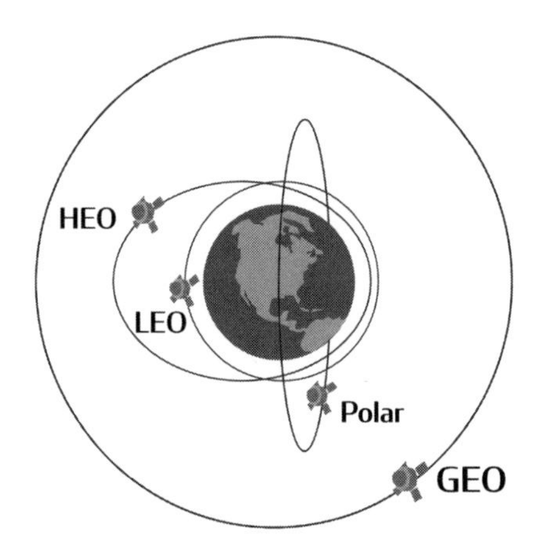

Satellites need to travel extremely fast to stay in orbit—an unimaginable thirty thousand kilometers per hour. That's eight kilometers every second, which is faster than a speeding bullet. If you want your satellite to reach a higher orbit or to travel to the moon or Mars, it has to go even faster. That's why we need rockets to reach space—they're the only machines that can fly that fast.

To help rockets get up to that speed, most launch toward the east. Why? To save fuel! When a rocket launches to the east, it gets an extra boost from the Earth's spin. At the equator, the ground is moving along at 1,600 kilometers per hour around the center of the planet. That's free energy the rocket can take advantage of. If it launches in the opposite direction, west, it has to burn extra fuel to counteract the spin of the Earth.

If you want to see the entire surface of the Earth, not just the part around the middle, you have to send your satellite into a polar orbit, which means it will pass over both the North and South Poles. It takes a satellite about an hour and a half to circle the Earth once, so each time it comes around, the Earth has turned a little bit underneath it. Over the course of an entire day, the satellite will see the entire surface of the planet.

It's hard to believe, but it's even possible to make a satellite stay still in the sky and not seem to move at all. To do that, it has to be in a very special orbit called geostationary (GEO). This orbit occurs when a satellite is sent really, really high—thirty-six thousand kilometers above the Earth's equator. At this orbit, the satellite will take twenty-four hours to circle the Earth. Of course, we know that twenty-four hours is also the time it takes the Earth to rotate once. So if the satellite is taking the same time to circle the Earth as the Earth takes to turn itself, the satellite will remain over one spot as both it and the Earth rotate at the same speed. It's like turning your body around with your hand held out in front of your face—your body and hand are turning at the same rate, so your hand is always in front of your nose. This is why satellite dishes that you see in backyards or on top of buildings always point to the same place in the sky. From the ground, it looks like the satellite is still, but in fact it's whizzing through space at thousands of kilometers per hour.

And if you like having your picture taken, look up on a clear day and wave. Some satellites up there may be spying on you. Spy satellites are used by the military so that one country can find out what another country is doing. Are they building rockets or weapons? The spy-satellite images will reveal the answers.

Some spy satellites want to see without being seen, so they are placed in an HEO—highly elliptical orbit—which is both high and low. The orbit is football-shaped, with one end way out in space, the other close to the Earth. When the satellite approaches the highest point of the orbit, it slows down and hovers over one spot for a while as it watches what's happening on the ground. If it sees something interesting and needs a closer look, it swoops down to the lowest point in its orbit. It whizzes by, secretly snaps close-up pictures, then retreats into deep space before anyone notices.

If you want to see a working satellite, turn on the Weather Channel or go to a weather website. There, you'll see images of clouds and storms as they appear from space. Weather forecasters depend on satellite images to see where storms are developing and how fast they're moving. That's how they can tell you when bad weather is likely to reach your area. Hundreds of satellites orbit the Earth every day, sending back pictures of the weather so that we know what to wear when we step outside.

Satellites can do a lot more than just help us predict the weather, though. They can also tell us a lot about environmental problems.

The hole in the ozone layer is an example of this. The ozone layer is a level in our atmosphere that protects us from harmful radiation from the sun. You can't see the holes in it with your eyes, but thanks to satellites, we can see that this natural sunscreen is disappearing over the North and South Poles. People didn't even know about the ozone holes before the 1970s. Now, thanks to satellites, we can watch them grow and shrink every day.

The same is true of El Niño, a huge blob of warm water that moves back and forth across the Pacific Ocean. When El Niño comes to North America, it disturbs weather patterns along the West Coast. It's only from the high perspective of space that we can see the shape of these huge systems in the oceans and the atmosphere.

Satellites also play a very important role in helping us understand changes on the Earth's surface that may be the result of climate change. Looking at satellite images taken over the decades, you can see recession in the tree cover in the Brazilian rainforest, the declining amount of snow that's on the Earth, and how the ice in the Arctic Ocean is gradually disappearing. Satellites allow us to learn about these important changes and see our world in incredible ways, but it's up to us to make sure we put that knowledge to good use.

14

How Do Telescopes Work?

Telescopes are big magnifying glasses that make faraway things look closer. They come in different sizes, from small instruments you can hold in your hand to giant structures housed in huge domes on mountaintops.

The first telescope was invented by a Dutch lens maker named Hans Lippershey. His day job was making reading glasses. But one day, so the story goes, he was checking the clarity of two lenses by holding them up to a window. For whatever reason, he held one lens in front of the other and looked through both. To his surprise, he saw a magnified image of a church steeple across town. Oddly, the lenses also turned the steeple upside down, but Hans realized that the two lenses working together could bring distant objects closer.

So the telescope was born. Early models were long tubes with a lens at either end like the ones pirates use in the movies. They were used mostly

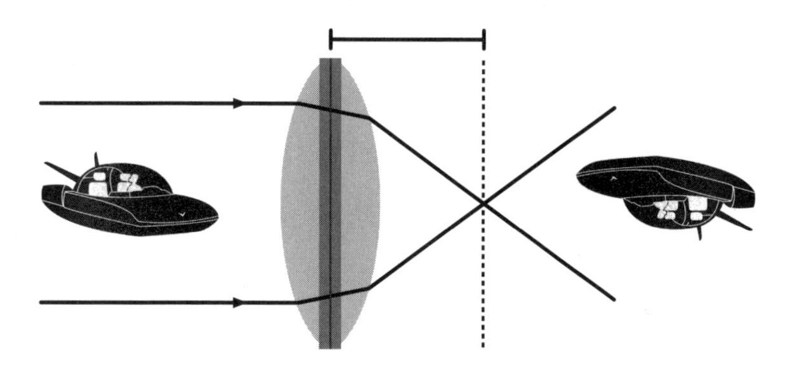

on ships to spot the flags of other ships in the distance to tell if they were friend or foe, or to see lighthouses along the shore.

Later, the Italian astronomer Galileo pointed a telescope at the moon and was amazed to see mountains and valleys on its surface. Then he used it to look at Jupiter and saw four little moons orbiting around it. He even saw the rings of Saturn. Telescopes, then, were pivotal to the creation of modern astronomy.

We've come a long way since those early models. Today, there are two types. Some use the old design, with glass lenses mounted in a long tube. One drawback of those types of telescopes is that they can't be too big because glass is heavy and actually sags under its own weight. When that happens, the lens loses its shape, so the light passing through it is distorted and the image becomes blurry. The largest telescope of this type has a lens at the front that is one hundred centimeters, or one meter, across.

In astronomy, bigger is better. The bigger the eye on the sky, the farther into space we can see. So another way to capture more light is to use a curved mirror. (Some mirrors that you can hold in your hand have two sides, one flat and one curved—the curved side makes your face look bigger.) A mirror telescope—also known as a reflecting telescope—can be made much larger than one using a lens because the light bounces off the shiny surface instead of going through it. That means that the back side of the mirror can be supported by strong beams and girders that hold the mirror's shape and make the telescope look more like a bridge than a scientific instrument.

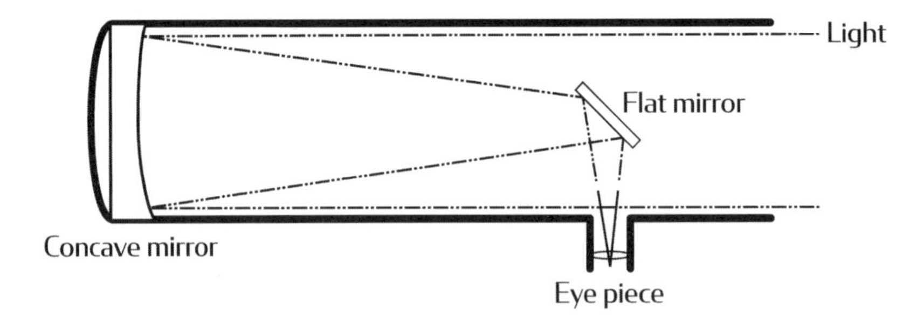

ON THE DRAWING BOARD

A number of new, super-giant telescopes are under construction around the world. As each one is completed, it will be the largest in the world, only to be overtaken by the one after it, which will be even larger. How large do you think they can get?

GIANT MAGELLAN TELESCOPE

This is actually seven telescopes all mounted on one frame, resembling a giant flower. Each mirror is eight meters across, providing a total span of 24.5 meters. It will operate from a mountaintop in Chile's Atacama Desert, one of the driest places on Earth, providing many clear nights throughout the year.

EXTREMELY LARGE TELESCOPE

This telescope will have a mirror thirty-nine meters across, made up of almost eight hundred segments, each 1.4 meters wide and only fifty-five millimeters thick. The giant instrument will rest in a dome the size of a football stadium. It will also be located in Chile.

Reflecting telescopes are the largest in the world. Some have mirrors that weigh many tons and span more than ten meters. A larger mirror captures more light, which allows them to see farther out into the universe. But even reflecting telescopes run into a problem when they get really big. Glass is heavy, so as the mirror grows, so does the structure that supports it.

The Hale Telescope in California, which was the largest in the world for many years, has a five-meter mirror that weighs thirteen tons. All of that weight requires a 481-ton structure of giant beams and tubes to support it. Not only that, but all of that weight must be perfectly balanced so the telescope can swing around with the precision of a fine watch to point at any part of the sky.

Today, telescopes are more than twice as big as the Hale. Scientists and engineers get around the weight problem by making the mirrors out of many smaller, thinner, and lighter sections that fit together like a jigsaw puzzle. In the future, these giants on the ground will grow to enormous sizes, with mirrors spanning more than thirty meters across.

Of course, we don't just want to get a quick glimpse of our universe—we want to be able to preserve it in pictures. When big telescopes were first built, astronomers took photographs of the stars and planets using cameras that were operated by hand. That meant spending long, cold nights on top of mountains, where telescopes had the best views, and sometimes, on the really big instruments, climbing up to the top of the structure to special cages where the cameras were located, and riding in the telescope itself to take the pictures.

Today, most telescopes are fitted with digital cameras and other instruments that can be operated by computer from a warm, comfortable office anywhere in the world. Usually, the only people up on the mountain are technicians who make sure the instruments are working properly. In many cases, astronomers don't even visit the telescopes they are using. But thanks to the amazing views of the universe our telescopes provide, astronomers can still visit the stars!

SPACE PLACES

Here are some places you can visit to see big telescopes where astronomers study the universe:

DOMINION ASTROPHYSICAL OBSERVATORY

Located in Victoria, British Columbia, this was the world's largest operating telescope just over one hundred years ago. You can still visit it today for star parties on Saturday nights, where you watch the big telescope in action and look through the smaller ones owned by members of the Royal Astronomical Society of Canada.

DOMINION RADIO ASTROPHYSICAL OBSERVATORY

Stars give off light, which we see with our eyes, but they also give off invisible radio waves, especially when they explode or when galaxies collide with each other. Also located in British Columbia, this observatory features a giant radio dish that detects signals from exploding stars and some of those other, more violent events in the universe.

DAVID DUNLAP OBSERVATORY

This big telescope, just north of Toronto, Ontario, is open to the public on weekends. It is the largest telescope in Canada operated entirely by volunteers from the Royal Astronomical Society, and it's only an hour from Canada's largest city.

JASPER DARK SKY FESTIVAL

Every year, hundreds of sky watchers gather in Jasper, Alberta, for the largest star party in Canada. It features outdoor festivities, special presentations, storytelling, and many, many telescopes to look through under the wonderfully dark skies of Jasper National Park.

MOUNT WILSON OBSERVATORY

This is the very telescope, located in California, that Edwin Hubble used to discover the expanding universe.

PALOMAR OBSERVATORY

For many years, this was the world's largest telescope, with a mirror 5.1 meters (two hundred inches) across. It's still the largest telescope in North America. The gigantic structure weighs 480 tons, and the whole thing rests under a huge dome that's forty-one meters (135 feet) tall.

MAUNA KEA OBSERVATORIES

On top of this extinct volcano, whose peak is a dizzying altitude of 4,200 meters above sea level in Hawaii, is the largest gathering of telescopes in the world. More than a dozen instruments, operated by many different countries, take advantage of the high-mountain perch above the clouds to get the clearest skies anywhere on the planet.

LOWELL OBSERVATORY

This telescope in Arizona was built by astronomer Percival Lowell, who believed that there were canals on Mars, possibly built by a Martian civilization. Neither canals nor Martians were ever found on Mars, but Lowell's idea spawned the notion of invasions from Mars that have become so popular in science fiction movies. This historic observatory, with its wooden dome, is named after him, and it's also the place where Pluto was discovered.

KITT PEAK NATIONAL OBSERVATORY

On top of this Arizona mountain are telescopes that look at the stars as well as the world's largest and most unusual-looking solar telescope, used to study the sun. It is one of the few places where astronomy happens during the day! Here, astronomers get a close-up view of the violent face of our home star.

YOU TRY IT!

Cardboard Galileo

With just a few household items, you can make your own refracting telescope, the same type used by Galileo more than four hundred years ago.

WHAT YOU NEED

1. Two magnifying glasses with handles
2. A cardboard tube (the ends should be as large as the magnifying lenses)
3. A sharp knife with a sawtooth edge
4. Measuring tape or a ruler
5. A pencil
6. Tape
7. A book

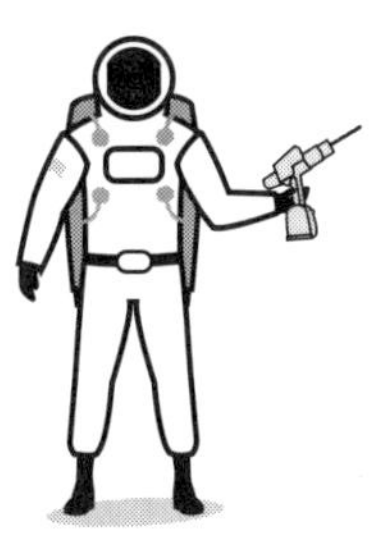

WHAT TO DO

1. Stand the book on its end on a table.
2. Hold up both magnifying glasses, one in each hand, and look through both of them at the book. Move the magnifying glass closest to you forward or backward until the words on the cover come into focus.
3. Measure the distance between the two lenses.
4. Mark two points on the cardboard tube the same distance apart from each other as the two lenses. Draw a straight line on the side of the cardboard tube connecting them.
5. Cut a notch into the cardboard tube at each endpoint. The notch should be wide enough to fit and hold the magnifying glass.

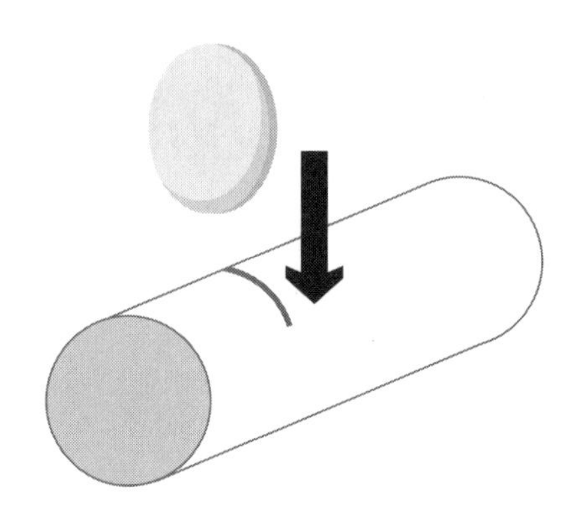

6. Cut along the line connecting the two points so that you have a slot running along the length of the tube.
7. Slide one of the magnifying glasses into the notch at one end so the handle is sticking out. Do the same with the other magnifying glass at the other end.
8. Hold the tube up to your eye using the handles of the magnifying glasses and look at a distant object. Slide the lens closest to your eye along the lengthwise slot until the object you're looking at comes into focus. You may have to cut more of the slot to find exactly the right spot.

Congratulations! You have just made a telescope. Use it to look at distant buildings and birds, and at night try looking at the moon. You will be amazed!

15

What's So Special About the Hubble Space Telescope?

The Hubble Space Telescope is probably the most famous telescope in the world, thanks to the perfectly clear, colorful images it has captured of objects all the way out to the edge of the universe. While it is not the largest telescope ever built, nor was it the first astronomical telescope sent into space, it was the largest to be sent into orbit, high above the turbulent, cloudy, and polluted air of the Earth's atmosphere. That gives it the clearest view of the stars possible. And because of its high perspective, it can look into space anytime, unlike telescopes on the ground, which can only look up on clear nights. In space, the skies are dark and clear all the time!

The telescope is named after Edwin Hubble, a famous astronomer who discovered that our galaxy, the Milky Way, is only one of billions of

others like it scattered across the universe. Not only that, Hubble also found that all the other galaxies are moving away from us, a sign that the universe is expanding.

Way back in the 1920s, Hubble began working at the Mount Wilson Observatory in California. He used the observatory's one-hundred-inch reflector telescope, which was the largest telescope in the world at that time. When Hubble (the person) peered out at fuzzy spots in the sky that looked like clouds in space, he discovered that they were distant galaxies. But there was something strange about their light. They appeared to be redder than they should be.

When an object out in space, such as a star or a galaxy, is moving toward us, the light waves are squeezed together a bit, making the light look blue. And when the object is moving away from us, the light waves are stretched out, so they appear red. The color of a galaxy, then, tells us whether it's moving toward or away from us.

The same thing happens when a train or car is passing by with its horn blasting. When the vehicle is approaching you, the sound waves are squeezed together, so the horn sounds higher. When it passes by, the horn becomes lower because the sound waves are stretched out. The faster the vehicle moves, the bigger the change in sound. When you are inside the vehicle, though, the sound of the horn does not change because you are moving with it.

Hubble was astonished to discover that the light waves from nearly all the galaxies he observed were red. And it didn't matter what part of the sky Hubble looked at: up, down . . . they were all "red shifted," which meant they were moving away from us. It looked like *we* were at the center of the universe. Not only that, but he saw that the galaxies farthest away from us were moving the fastest.

It all seemed incredibly strange. If all those galaxies were rushing away from us, that means they must have been closer together in the past. Logically, if we could run time backward and travel into the past, the galaxies would get closer and closer together. So it seemed the universe used to be smaller and everything was once packed into a tiny space. If we could go all the way back to the very beginning, the whole universe would have been squeezed into a single ball of hot energy . . . a powerful bomb waiting to explode.

And explode it did. We came out of a big explosion, something scientists call the Big Bang. So not only does it appear that we were once at the center of the universe, but we were also in the middle of a big explosion.

We think of explosions like fireworks, where everything bursts out of one spot and flies off in all directions, with the sparks traveling at the same speed away from the explosion site. But the Big Bang was the explosion of the universe itself. And since we live in the universe, we see the explosion from the inside out, which is why everything seems to be rushing away from us in all directions.

Here is the really hard part to wrap your head around. The expanding universe is unlike fireworks in another way. The universe is not filling up a large, empty space the way fireworks fill the sky. The universe is expanding because space itself is getting larger.

It's tricky to think of empty space getting bigger, but that's what Edwin Hubble saw when he looked at galaxies in space. Those farthest away were expanding faster than those that were closer, and that can only happen if space is stretching the way a balloon does when you blow it up: space expands and makes galaxies move away from each other, even though the galaxies themselves are not moving through space.

Picture our universe as a deflated balloon. Imagine you've drawn dots all over it to represent planets and galaxies. Now imagine blowing it up. The planets and galaxies stretch away from one another as you blow, but they don't change where they are on the balloon.

Now picture a very tiny version of yourself standing on one of those dots. No matter where you were standing on that balloon, it would always look like you were standing still and all the other dots are moving away from you. It doesn't matter what dot you stand on, the effect will be the same. What does this mean? It means that as far as we know, there is no center of the universe.

So is there an end? Is there an edge to the universe? When you're in a big crowd, surrounded on all sides by masses of people, you can't see where the crowd ends, right? The same goes for our universe. We're inside it, so we don't see an edge. We don't even know if there is an edge. It's not like we can step outside the universe to take a look.

The Hubble Space Telescope has proven that Hubble was correct—the universe is expanding. The telescope has even measured how fast it's happening. It turns out, not only is our universe expanding, but the rate of expansion is speeding up! That means that it will never stop expanding. Eventually, many billions of years from now, all the galaxies will be so far away from us that we will no longer be able to see them. So get out and enjoy the night sky while everything is still within sight!

YOU TRY IT!

Make a Rubbery Universe

When we think of objects getting farther away from us, they are usually moving through space. But the expanding universe is space itself between the galaxies that is getting bigger, which is hard to imagine. A rubber band helps to show how the effect that Hubble saw actually works.

WHAT YOU NEED

1. One thick rubber band
2. A marker

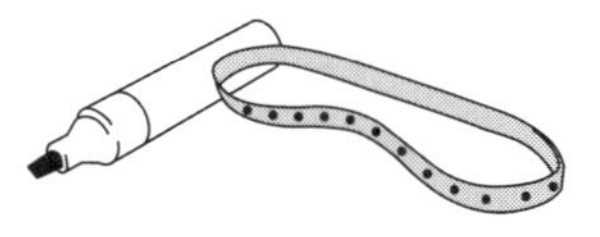

WHAT TO DO

1. Make a series of dots along the rubber band that are all about the same distance apart. Hold the rubber band by the ends and stretch it to see what happens to the dots. The dots are getting farther apart because the space between them is stretching. But the dots are not actually moving across the rubber. Rather, they are being carried along as it stretches. The same thing is happening in the universe. Space itself is expanding and galaxies are getting farther apart.
2. Rest one hand on a table and hold one end of the rubber band still. Imagine being on the dot closest to your table hand while looking along at the other dots. Now stretch the rubber away from your hand on the table. Notice how the dot closest to the stationary end does not move as much as the ones farther away, and the closer you get to the moving end, the faster the dots are moving. Astronomers see the same thing happening in our universe.

Galaxies that are the farthest away are moving the fastest. This can be seen in all directions in the sky, so it looks like we are in the center.

3. Now rest your other hand on the table to hold the opposite end of the rubber band still. Imagine yourself on the dot closest to that hand while you stretch the rubber. Now all the other dots seems to be racing away in the opposite direction, but it still looks like you are at the center. In fact, it doesn't matter what dot you choose; if you imagine standing on it as the elastic stretches, it will always seem like all the other dots are moving away from you and you are standing still.
4. Finally, stretch both ends of the rubber band at the same time and make the center dot remain stationary. If you were on the middle dot, you would still see galaxies moving away from you and think you are in the center.

In all of the methods outlined above, you were simply stretching a rubber band. No matter what dot you imagined yourself on, you would think you were in the middle of the rubber band universe. That's one reason no one really knows where the middle of our universe is. We think we're at the center, and no matter where you are in the universe, your perspective would make it appear as though that was exactly the case. The truth? If everywhere is the center, then there probably is no center!

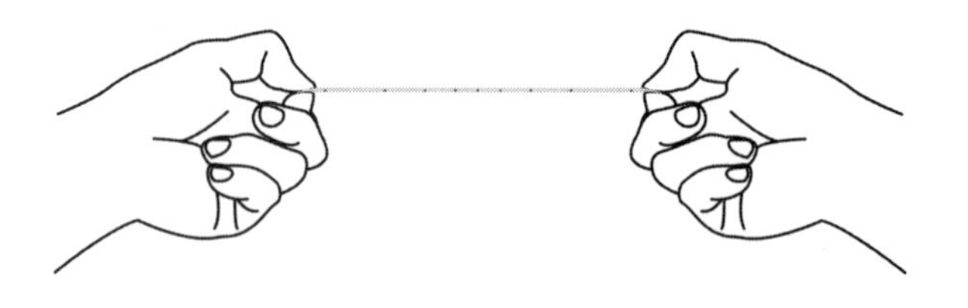

PART 3

Fly Me to the Moon . . . and Beyond

Humans in Space

16

How Do I Become an Astronaut?

Astronauts are very special people. They have a good education, they're strong like athletes, and they go through years of training before they fly in space. In the early days of space exploration, all astronauts were military test pilots. That's because no one knew what it would be like to leave the Earth, and the rockets taking them there had never been flown before. Test pilots train to fly new types of aircraft and are used to putting themselves into dangerous situations. So they were considered the best people to become astronauts.

But now, more than sixty years later, almost anyone can become an astronaut . . . if they possess the right qualities.

STEP ONE: GET AN EDUCATION

No high school dropouts have made it to space. An education in science, engineering, medical, or technical subjects will prepare you with the

knowledge needed to fly in space. Astronauts have to know about the technology of rockets and space stations because if something breaks in space, it has to be fixed in space. You can't call a plumber to repair a space toilet. Becoming a medical doctor makes you an excellent astronaut candidate because doctors are needed to treat anyone who gets injured or sick during a mission.

Then there is the science you will be doing once you leave the Earth. There are all sorts of things being measured in experiments in space—the health effects of spaceflight on the human body, how plants grow without the effect of gravity, or the chemistry of combustion and how things burn in weightlessness. Some of those experiments, such as environmental studies of the Earth, look down at the planet from high above, while others use instruments to look out to the edge of the universe. People from many different science backgrounds are needed to accomplish it all.

Finally, now that astronauts from many different countries fly to space together, you will likely have to learn another language.

STEP TWO: BE PHYSICALLY FIT

No one who is overweight or out of shape has crawled into a spaceship and blasted off the Earth. Spaceflight is hard on the body, so you have to be strong to survive it. During launch, you feel three times your weight from the acceleration of the rocket. Then, as soon as the engines shut down, you are thrown into instant weightlessness, which gives you a sense of falling that can turn your stomach. If you remain in space for many months, your body becomes weaker because your muscles don't have to work as hard when you don't weigh anything. And after all that, you have to handle the gravity of Earth after you return. All of this takes strength and stamina. You don't have to be an Olympic athlete, but you do have to be very fit and exercise often.

STEP THREE: BE A TEAM PLAYER

No one flies in space alone. It takes a team of people to fly in space and another even larger team on the ground to keep them safely up there. If you are the type of person who prefers to do everything yourself or who doesn't like to work with others, you won't be chosen to become an astronaut. Space is a very dangerous place, so astronauts need to look out for one another. If a task becomes too difficult or something goes wrong, you have to be able to ask for assistance and know that your crewmates will be there to help.

If you're one of the lucky few who is chosen to become an astronaut, you'll eventually end up training at the Johnson Space Center in Houston, Texas, and the Yuri A. Gagarin Research & Test Cosmonaut Training Center in Moscow, Russia. All astronauts go to these centers to learn how to fly spacecraft, do space walks, and operate the equipment on the International Space Station. The experience at those centers is as close to actually flying in space as you can possibly get.

It takes at least two years of basic training to become an astronaut. A lot of that involves time in classrooms, where you learn about the science of spaceflight, orbits, how to rendezvous with other spacecraft such as the International Space Station, and the details of reentry into the Earth's atmosphere. Then it takes several more years of training while you wait to be assigned to a mission. During that time you will learn how to do space walks in a giant swimming pool and help with missions already in space by becoming the "CAPCOM," or capsule communicator, who speaks directly with the astronauts in orbit.

In one building in Houston, there's a full-size replica of the International Space Station, where you can step inside the different modules, see how the equipment is laid out, and practice all the activities you'll be doing in space. You'll be taken up in jet aircraft to experience the forces that are felt during rocket launches, and you'll work with teams of other astronauts, cosmonauts, and ground engineers to make it all happen.

When a rocket blasts off, your body is squeezed by the sudden acceleration, the same way you are pushed into the seat of a car when it accelerates quickly at a green light. But rockets take off much faster than cars, going from zero to the speed of sound in about a minute and up to thirty

SPACE PLACES

Here are some centers where you can get a taste of what it is like to become an astronaut.

SPACE CAMP

Located in Huntsville, Alabama, and home of the first American rocket scientists, this camp has full-size simulators of the space shuttle and space station. After you climb in, you can feel what it's like into launch into space, walk on the moon, and work in zero gravity—or even operate a console in mission control during a mission.

CAMP KENNEDY SPACE CENTER

This camp is located at the Kennedy Space Center in Florida, where real rockets launch into space. Besides all the activities of other space camps, you will have a chance to meet an astronaut, tour the genuine NASA launch facilities, and, if you time your visit right, maybe see a rocket launch for real.

COSMODOME

This space center in Laval, Quebec, holds artifacts from the Canadian space program as well as spaceflight simulators and a space camp.

LYNDON B. JOHNSON SPACE CENTER

Real astronauts go to Houston to learn how to fly in space. It is also mission control for the International Space Station. You can take a tour of the buildings where astronauts train, see a real moon rocket, and fly simulators at the visitor center.

thousand kilometers per hour in only eight minutes, so the force you feel on liftoff is a lot more than what you'd experience in a car.

One device used to train for the stresses of launch is a centrifuge, a long swing arm with a capsule on the end. The seat inside the capsule has a four-point harness, like those in a fighter jet, that goes over your shoulders and across your thighs to keep you in tight. When the door is closed, you can't see out, so when the arm begins to smoothly swing around, you don't really feel like you are moving. What you do feel is your body pressing down into the seat as though you are gaining weight. As the speed increases, so too does the force on your body until you feel three times heavier than you would in a normal chair. That is the force astronauts feel during a launch into space. When a Russian Soyuz capsule returns to Earth and hits the atmosphere, the force, known as g-force, can go up to six or seven. Even at just three gs, the body feels heavy and you have to work harder to lift your arms over your head in order to flip switches. That is one reason astronauts exercise a lot. They need to be strong just to get off the Earth.

Everyone who flies in space must get used to the fact that their body, and everything else around them, has no weight. You can fly through the air like Superman or Supergirl with a simple push off a wall. And you can fly in any direction because up and down no longer exist, and anything you hold out in front of you will stay there when you let it go. Nothing falls to the floor. If you need to pass something to someone else, simply float it across the room.

It takes a little time to get used to floating all the time. First-time fliers tend to bump into other people and bounce off walls because they push too hard and have trouble controlling themselves. But they soon learn to do everything in slow motion and move using mostly their fingertips. No one wears shoes because no one walks. Everyone flies!

One way to experience weightlessness on Earth is in a swimming pool. If you are really good at holding your breath, you can go to the deep end of a pool. Take a deep, deep breath, then slowly let the air out of your lungs until your body begins to sink. Hold your breath at that point and try to suspend yourself in the middle of the water so you are halfway between the bottom and the surface.

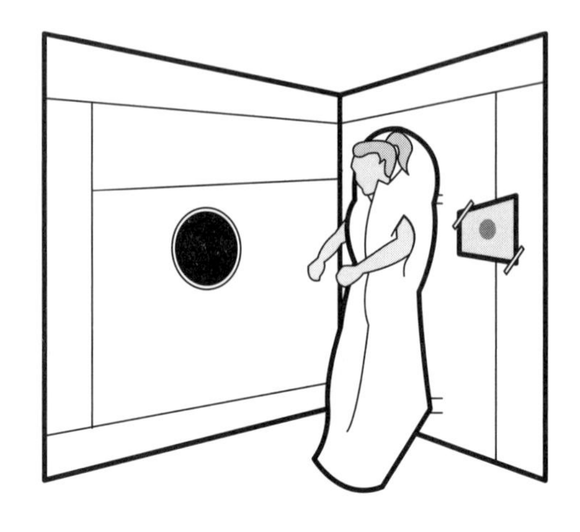

When you're floating still like that, neither sinking nor rising, you're what is called neutrally buoyant. You're experiencing what it is like to have nothing touching you anywhere on your body. That is what weightlessness feels like. It's like leaping off a diving board, then, just before you hit the water, the pool disappears and you just keep on falling and falling. It sounds scary, but for astronauts, it is a lot of fun.

At the end of the day, astronauts still have to go to sleep, just like the rest of us. But how can you lie on a bed if you don't weigh anything?

Simple. You hang yourself on the wall.

Sleeping quarters on the space station are about the size of a small closet. There is a laptop mounted on one wall so you can check messages from Earth, a small window to look out, and a sleeping bag attached to the opposite wall. You crawl into the bag and then you can either tuck your arms inside or stick them through holes in the side of the bag and let them float free in front of you. Some astronauts have slept floating in the middle of the space with their legs crossed and arms folded. Whatever the position, everyone finds sleeping in space extremely comfortable because there is no pressure on any part of their body from the bed, pillow, or covers.

Astronauts make spaceflight look easy, but some say that after returning to Earth, getting used to gravity again is just as confusing as going up in the first place. After living in a world where your body weighs nothing, the return of gravity feels cruel. Everything is heavy, your head feels like a bowling ball on top of your neck, and your arms feel like logs. One cosmonaut who spent more than a year in space said that even his eyelids felt heavy when he opened them. Some astronauts have trouble balancing when they get back—the organs in the body that tell us where up and down are get turned off in space, which makes it tricky to walk on Earth when you get back, especially when you're going around corners.

If all of this training to become an astronaut is not for you, there are thousands of other jobs that keep those astronauts up in space: engineers who build the spaceships and launchpads; technicians who work on space suits; doctors who monitor the health of astronauts; scientists who design experiments for space; and robotics people who work on the Canadarm and other systems.

While you're waiting to become an astronaut, try spotting their spacecraft from down here on the ground. The space station is so big it can be easily identified in the night sky. It looks like a very bright star or an airplane moving smoothly across the sky, except it doesn't have any blinking lights on it. You can see it from every major city in North America. You just have to know when it's passing over your city and where to look in the sky. There are a number of web pages that will tell you when the International Space Station is passing over your city or town. As you watch the station fly overhead, think about the fact that there are people living inside it. If you dream hard and work hard, perhaps one day you will be one of them.

YOU TRY IT!

Spin Cycle

Want to know what your first couple of days in space will feel like? Try this.

WHAT YOU NEED

1. Lots of floor space, or a wide space outside

WHAT TO DO

1. Bend over and tilt your head down as far as possible so you're looking straight down at the ground.
2. While holding your head in this position, spin yourself around as fast as possible at least five times (or more, if you can).
3. After the last spin, stand up and try to walk in a straight line. You'll probably find yourself staggering to one side.

What just happened? Any time you move your head, the fluid in your ear canals sloshes around, sending signals to your brain that your head is moving. When you stand up after twirling around, the fluids keep spinning for a few seconds, signaling to your brain that your head is tilting to the side when it's really not—your brain thinks you're falling over and tries to compensate, so you stagger until your brain receives the new message that you're balanced.

In space, the false signals to the brain make astronauts lose their sense of up and down. Thankfully, that feeling goes away as the brain learns to ignore the false signals from the body and relies just on the eyes to figure out what the body is doing.

17

Who Was First in Space?

When you think about traveling in space, you might picture the American space shuttles or maybe the giant *Apollo* rockets that went to the moon. But Russia is also a space-faring country. In fact, it was the first to get there.

More than one hundred years ago, Konstantin Tsiolkovsky, a Russian scientist and schoolteacher, came up with the idea of a rocket-powered spaceship that could carry people into space. His idea of how a rocket works is the founding principle of rocketry still used today.

Russian rockets have flown into space more often than any others and are still the most reliable way to reach space. Since the Americans retired their space shuttle fleet in 2011, they have been hitching rides on Russian rockets to get astronauts and supplies up to the International Space Station. The two countries help each other and work together. But it hasn't always been that way . . .

The beginning of the Space Age was actually a space race, and it was all about rockets. The United States and Russia, then called the Soviet Union, were both in a race to build a rocket big enough to get into space.

Both countries built bigger and bigger rockets, trying

to beat each other and prove who was more powerful by becoming the first country to successfully send a man-made object into orbit. Eventually, in 1957, the Russians won the race when they launched a small, shiny, spherical satellite about the size of a basketball. *Sputnik* became the first object other than the moon to orbit the Earth. It was our first artificial satellite and it marked the beginning of the Space Age.

Sputnik wasn't very big, but it did prove that the Russians had rockets powerful enough to reach space. On their third launch, they sent another probe, not much larger than *Sputnik*, past the moon. Then, to prove their rockets were even more powerful, they sent up a dog named Laika—the first living creature to fly in space. Unfortunately, she also became the first animal to die in space because the Russians didn't have a way to bring her capsule home. She was followed by two other dogs—Belka and Strelka—who did make it back to Earth and can now be seen stuffed in a museum in Moscow.

But there was still one more big, important step that the Russians and Americans wanted to take . . . and they each desperately wanted to be the first to take it. They wanted to put a human being into space.

In 1961, a young Russian pilot named Yuri Gagarin climbed to the top of a Vostok rocket and blasted into orbit. He was the first spaceman. The Russians called him a cosmonaut. He was an instant hero. The name Yuri Gagarin is now in history books around the world. Following Gagarin's flight, Valentina Tereshkova became the first woman in space—another Russian first.

The rockets that sent Gagarin and Tereshkova into space might look simple compared to a space shuttle, but the amazing thing is that the Russians still use a similar design to fly people into space today. Sure, the rockets are bigger, but Russian rockets have not changed much since the early days because Russians believe that if you get it right the first time, you don't mess with it.

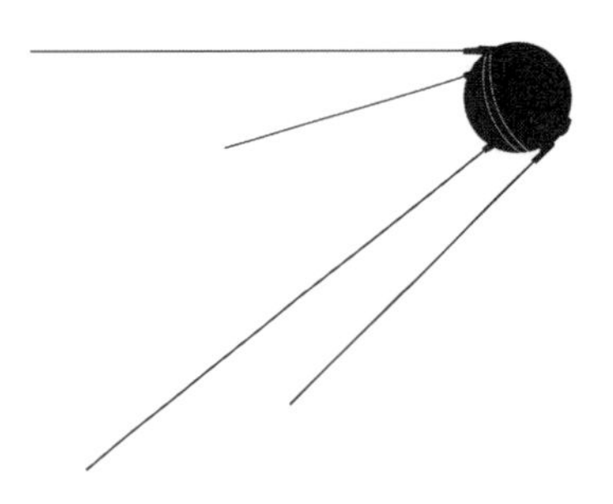

Today's Soyuz rockets have the same shape as their predecessors, and they still use small capsules—only now they carry three people

SPACE PLACES

If you want to fly in space "Russian style," it all starts at Star City outside Moscow. Star City is also known as the Yuri A. Gagarin Research and Test Cosmonaut Training Center. All Russian cosmonauts, including Gagarin himself, trained here. That's where they learned to fly the Soyuz capsule, which carries cosmonauts and astronauts into space and back.

A second place is the "Cosmos" Pavilion in Moscow. This huge building contains life-size models of Russian spacecraft past, present, and future. You can step inside the Russian section of the International Space Station and see their version of a space shuttle, called *Buran*, which flew in space only once.

instead of one. It is very cramped inside, with the three cosmonauts jammed in shoulder to shoulder. Fortunately, they don't have to spend very long inside because the Soyuz is used only as a taxi to get up to the space station and back down again.

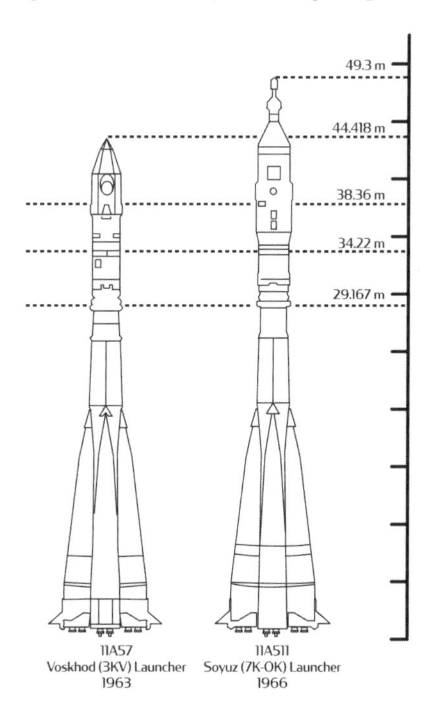

Landing in a Soyuz is still somewhat primitive. The capsule has no wings, so on reentry, it falls through the atmosphere like a meteor and then opens a parachute. Just before the capsule hits the ground, retrorockets fire, slowing it before it thumps and lands. There isn't much control, so if the round capsule happens to land on a hill, it rolls.

The one part of the space race that the Russians lost was the race to land a man on the moon. They did build a giant rocket called the

N1 that was designed to carry two men to the moon. But unfortunately, it exploded during the four times it was launched, and the program was canceled. So American astronauts were the first to plant a flag on another world. But that didn't mean the Russians were out of the space business.

While American astronauts were collecting rocks on one part of the moon, Russian robots were roaming on another. One of them, called *Lunokhod*, was the first remote-controlled car to be driven on another world. It was about the size of a bathtub.

Russian robots continued to make other firsts in space by visiting other planets. Their *Venera* robots are the only probes to have landed on Venus. Instead of sending people to the moon, the Russians became experts at building space stations in orbit around the Earth. They've built lots of them—seven successful altogether. The last and largest was called Mir.

In space, where everything is weightless and floats, the modules of the Mir space station were assembled like the branches of a tree, sticking out in many different directions and joined by hatches. It got very weird when you went from one room to another. When you floated through a door or hatch in the Mir space station, you had to change your idea of where up and down was, roll your body into that position, and go into the next room. Imagine going through a door to find that the floor and ceiling are in completely different positions from what you'd expect.

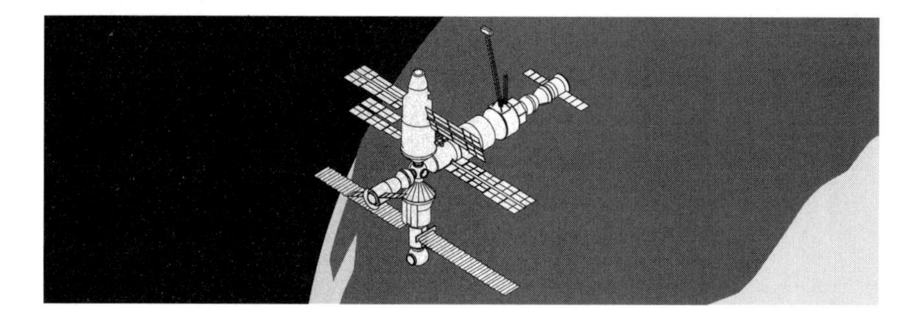

Russian cosmonauts who lived on Mir for extended periods got used to this, but astronauts from the space shuttle who came up to visit Mir for only a few days at a time found it very confusing. They had to tape red arrows on the walls inside Mir pointing the way back to the shuttle because they were getting lost in this maze. The new International Space Station has solved that problem by making sure all the modules have up and down in the same direction.

The Russians did make one attempt to build a space shuttle. They called it *Buran*, which means "snowstorm" in Russian. It's no coincidence that it looked similar to the American shuttle, because the Russians copied the American design, which had already been proven to work in space.

Buran flew in space only once, on a short test flight with no people on board. That one flight was just to prove that the Russian shuttles could fly, but since then, no others have left the ground.

Today, Russians and Americans work together in space, along with other countries, such as Canada, Japan, England, France, Germany, and Italy, who all got together to build the International Space Station. And there are two control centers for the Space Station—one in Houston, Texas, and one in Moscow, Russia. Both English and Russian are spoken in space.

In the future, if humans travel to Mars, it will likely be done through a similar international effort involving many countries because the cost of going to Mars is so astronomically high. No single country would want to pay so much to visit another planet. So, instead of the competitive space race that led us to the moon, a Mars trip will be accomplished through cooperation—we'll all get there at the same time, rather than trying to be there first and doing it alone.

YOU TRY IT!

Floating Freely

Spaceflight is actually falling all the way around the Earth without hitting it. That's all that weightlessness is—just falling without ever hitting the ground. On Earth, we can only experience this for a few seconds before we hit the ground, but here is a way you can show how all objects fall at the same speed, and why objects in space float in front of your face.

WHAT YOU NEED

1. A swimming pool with diving board
2. A ball

WHAT TO DO

1. Stand at the end of the diving board and hold the ball straight out in front of you.
2. Jump off the diving board as high as you can.
3. Let go of the ball while you are in the air (don't throw it).
4. You and the ball should hit the water at the same time.
5. You have just experienced weightlessness like an astronaut!

When anything is falling, it's weightless. Both you and the ball are weightless when you are in the air. And the neat thing about falling is that gravity makes all objects fall at the same speed.

When you see astronauts floating around inside their spaceship, it looks like there is no gravity in space. Actually, there is lots of gravity up there, almost as much as there is down here on the ground. The reason the astronauts are floating is because they're falling all the time.

So just like you and the ball hitting the water together, astronauts, their spaceships, and everything inside them is falling together around the Earth, which is why everything is weightless and why everything you do in space—moving from one side of a room to another, eating dinner, brushing your teeth—happens while floating freely.

18

What Happens to Your Body in Space?

Space is a lazy place. Your body floats all the time, so you don't have to walk. And anything you pick up floats along with you, so your muscles don't get much of a workout. It sounds wonderful to be able to fly everywhere you want to go, but that lazy environment causes your body to change in many different ways that are not healthy. And if you stay in space for too long, you may not be able to walk when you return to Earth.

When astronauts first arrive in space, they sometimes don't recognize themselves. The fluids inside their bodies that would normally be pulled down by gravity float up into their chests, necks, and faces. Astronauts find that their chests expand a bit and their legs become skinny—"bird legs," as they say. It also produces an effect called "puffy face." The face becomes rounder and puffy. The astronaut's eyes start to look squinty because of the swelling.

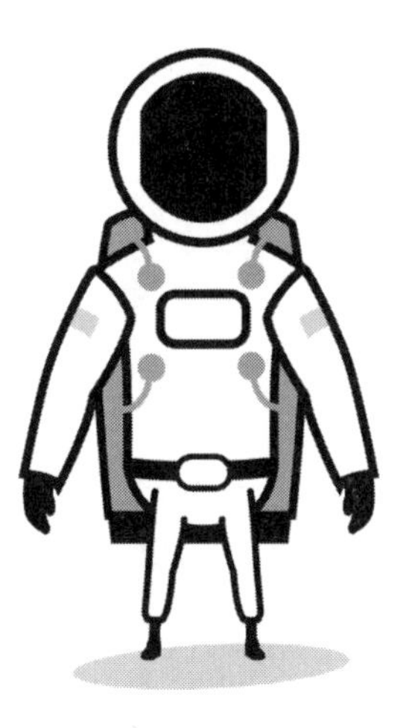

On Earth, gravity is always pulling the blood in our bodies down toward our feet. If it weren't for the pumping action of our hearts, we would all have really fat legs and big feet from all the blood flowing down there. We'd also be unconscious because our brains need blood to think. Thankfully, our heart is always working to pump blood upward, against gravity, so that it's circulated throughout the body to reach all the other parts, including our heads.

In space, that force of gravity is not present, so the blood is no longer pulled down toward the feet. The heart keeps pumping blood up toward head, but it remains there because there's no force to pull it back down. All that extra fluid in the head makes astronauts feel stuffy in the nose, like they have a cold. To counteract that effect, some astronauts take cold medicine while in space.

Another effect of the puffy face and chest is that the body thinks it has too much blood, so it starts to get rid of red blood cells. That's not a good thing because we need all of our blood to carry oxygen and other good stuff around the body. Our bones are affected in space as well. Our bones grow stronger the more we use them, and we use them all the time on Earth. Just standing up puts weight on your bones, allowing them to build strength as they hold you up. The more you walk, run, play sports, carry heavy things, or work out, the stronger and thicker your bones will become. Unfortunately, in space, where the body is weightless, our skeletons feel almost no force at all. When bones are not used, they become thinner. That's a problem for astronauts because they need those bones when they return to Earth. In fact, when astronauts return home, they have to take it easy at first so they don't break their bones while they regenerate.

To help prevent muscle and bone loss, astronauts living on the International Space Station have to exercise for two hours every day. They strap themselves onto a stationary bicycle, use bungie cords to hold themselves down to run on a treadmill, or use a spring device that simulates lifting weights.

If you've ever wanted to be a little taller, though, space flight might be for you. Your spine is a long chain of bones separated by soft discs. These discs act as shock absorbers so that walking on Earth is not so rough. In space, there's no downward force on your spine, so the spongy discs expand, pushing the bones of the spine apart. And what does that mean? It means you get taller by several centimeters. While that sounds great in theory, the expansion also pulls on nerves in the spine, which can cause uncomfortable back pain.

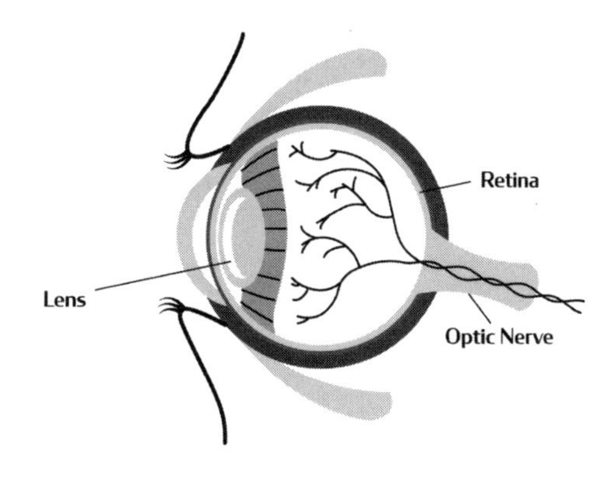

You don't need to go to the space station to find your space height, though. Try standing against a wall and have your friend place a pencil flat across the top of your head to mark how tall you are. Then lie down on the floor and have your friend make a mark right under your heel and another at the top of your head. Measure the difference between the two marks on the floor and compare it to your height on the wall. The floor marks should be slightly farther apart because your spine expanded slightly when you lay down.

Another body part affected in space is the eye. Some space farers see more clearly in space, especially those who need glasses on Earth. For others, their eyesight gets worse. Our eyes work in a way similar to cameras. Our eyeballs are shaped like round balls and are filled with a clear liquid. A lens on the front lets light in and makes an image on a screen, called the retina, located on the backside of the eye. The retina sends that image up to the brain through a nerve that comes out of the back of the eyeball. This whole system works together so that we can see.

The eyeball is not always perfectly round, though. Some people's eyes are a little less spherical, so the lens does not make a clear image on the retina and they see the world as a blur. Glasses or contact lenses in front of the eye can correct the issue by focusing the light in the right way.

In the weightlessness of space, eyeballs that aren't perfectly spherical sometimes become more rounded. The lenses focus in the right spot, which means those happy astronauts can put their glasses away for a while. Others, especially those who stay up in space for half a year at a time, find their vision gets worse and stays that way after they return to Earth. Doctors have found that extra fluid gathers in the brain and puts pressure on the eyes in space. In some cases, the optic nerve—the power cable that comes

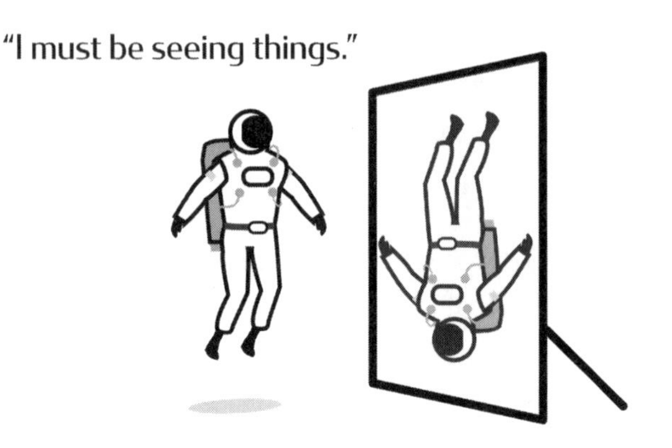

out of the eye and sends the image signal to the brain—becomes squeezed and bent from the pressure, so the signal to the brain is not as good. So far, there's no way of knowing who will experience better or worse vision in space.

If you are the type of person who feels queasy on a roller coaster, gets seasick on a boat, or doesn't feel well while riding in the back seat of a car, you may not enjoy flying in space, at least for the first day or two.

Motion sickness happens when our brain gets confused about what is happening to the body. Typically, the brain uses a variety of mechanisms to tell whether we are standing, sitting, moving forward or backward, turning left or right, spinning, or falling. The main input is through the eyes, which can see the world moving when we move.

But the brain also uses sensors in our skin—located on the bottoms of our feet, our hands, and our butts—so that we can feel pressure coming from different directions. Our ears act as important sensors, too. Behind our eardrums are little tubes called semicircular canals that are filled with liquid. When we move our heads, these liquids slosh around to tell the brain that the head is moving. It's those liquid movements in our ears that help us keep our balance when we walk.

On Earth, all of these sensory inputs usually agree with each other. What the eyes see, the body feels, and that gives us a pretty good idea of where we are going. In space, the eyes still work, but most of the other sensors shut down. The body is floating all the time, so there's very little pressure on the skin. The fluids in the ears swirl around in all directions, telling the brain that the body is falling, which it actually is (recall that weightlessness is just falling all the way around the Earth without hitting it). As a result, our brains in space get mixed messages. The eyes are saying the body is not moving, while the body is saying it is but doesn't know in which direction. Whenever the brain gets confused, we react by feeling dizzy. And the brain, wondering what's wrong, assumes the problems are due to some bad food you ate. It makes you sick, emptying your stomach just to rule out that possibility. About half of the people who fly in space feel nauseated for their first day or two.

It doesn't help that our digestive system is meant for Earth and counts on gravity to make it work. When we eat food or take a drink, we also swallow a bit of air. This accumulates in our stomach and forms bubbles, which rise to the top of the stomach and travel up our throats in the form of burps. Later, as our digestive system breaks down the food, bacteria that live in our gut produce other gases, such as methane, which accumulate in our lower bowel, form bubbles, and burst out of the other end.

In space, these bubbles still form, but because there's no downward pull of gravity, the bubbles don't rise to our mouths. Instead, they remain suspended in the stomach or the intestines, giving a feeling of "foaminess" in the stomach. Astronauts feel like they want to burp, but it doesn't happen, and if they do manage to get something up, let's just say it's usually pretty wet.

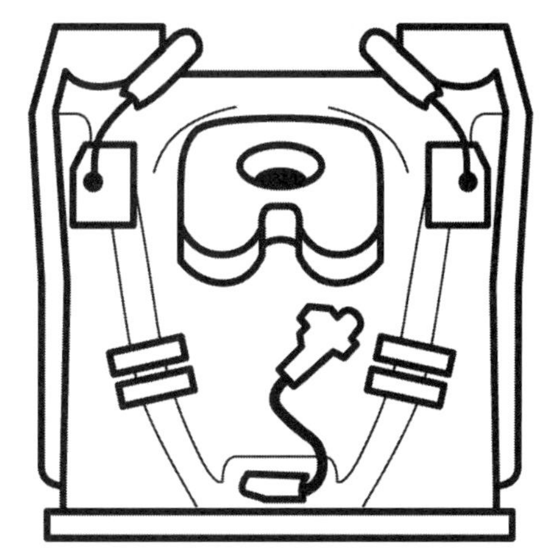

Finally, there's something we all have to do that no one likes to talk about, but I'm going to do it anyhow. How do you go to the bathroom in space?

A space toilet looks a lot like its counterpart on Earth, with a few important differences. Two bars swing out across your thighs to keep you from floating off the seat, which is soft so that it feels comfortable and conforms to your body. There are also foot restraints so you really feel like you are sitting—remember, people don't sit in space, they just float all the time, even while eating meals. But going to the toilet is something we like to do while sitting, so the space toilet is designed to make it feel like home.

Inside the toilet is a fan that draws air in under the seat and out through the bottom to act as a flush instead of water. The air flow makes sure that whatever you deposit in the toilet goes in the right direction. If all you need to do is pee, there is a suction hose with a cup on the end that is custom made for each crew member.

When everything is done, you use hand wipes to clean up, then the toilet lid is closed and the waste material inside is exposed to the vacuum of space, which freeze-dries it and kills all the bacteria.

That covers just a few of the unusual things that happen to the human body in outer space. But the truth is, we're still learning about the effects of zero gravity on our bodies, and as we learn more, we'll no doubt make further discoveries that alter how we conduct space travel.

One thing we know for sure is that flying in space is wonderful, but it takes astronauts away from their families and away from nature. Life in space is life spent in a tin can. Even if you're lucky enough to go on a space walk, you have to wear a bulky space suit, so you are not really outside or in nature. On very long journeys, astronauts begin to miss their family and friends, the smell of flowers, and the feeling of wind and rain on their faces. Fortunately, they now have access to the internet, so they can make regular calls to home.

If you're a person who can stay strong by exercising every day, doesn't get dizzy while moving in all directions, is willing to risk poor eyesight, doesn't mind having a foamy stomach, and is okay being away from friends and family for long periods of time, you could be astronaut material!

YOU TRY IT!

Space Face

Put on your space face and find out what you will look like when you leave Earth.

WHAT YOU NEED

1. A handheld mirror
2. A phone camera
3. Floor space near a wall
4. A friend to help

WHAT TO DO

1. Using the phone camera, take two close-up selfies showing just your face. Smile in one of them and keep a straight face in the other.
2. Have your friend help you stand on your head with your back and feet up against the wall for support.
3. While you are on your head, ask your friend to hold the mirror in front of your face so you can see yourself. Notice how different you look!
4. Have your friend turn the phone camera upside down and take two close-ups of your face, one smiling and one straight.
5. Compare how your face looks when you are standing upright and when you are on your head.
6. You now have your space face on!

When you turn upside down, the fluids in your body shift toward your head, making your face puffy and pink. Everyone—from Yuri

Gagarin, the first person in space, to people living on the International Space Station today—looks different in orbit than they do on the ground. The force of gravity no longer keeps fluids down into the lower body, so the face becomes rounder. Astronauts are sometimes surprised at how different they look when they catch their reflection in a mirror while in space.

19

How Do You Get Around in Space?

Have you ever wanted to ride a rocket? Getting around in space is very different from getting around here on the ground. You can't go anywhere in a straight line because everything in space is moving in circles. And it's moving really fast!

The first part of getting around in space is simply getting *to* space. Getting there means riding a rocket, the fastest vehicle there is. Why can rockets fly us through outer space but airplanes can't? The biggest reason is that there is no air in space, and airplanes just don't work when there is no air. There's nothing for the wings to fly through, and no oxygen for the engines to burn. Airplanes only get you so far.

Rockets work using a simple principle called action and reaction. Hot gases burn in the engine and blow out the bottom. That's action. The rocket moves in the opposite direction, which is the reaction. Gases go down, rocket goes up. It's basically a controlled explosion.

The only problem is that gases are very light and rockets are very heavy, so it takes a lot of gas to move the rocket. That's why they need so much fuel. In fact, most of a rocket is fuel. The part of the rocket that actually makes it to space is within the small nose cone at the top. The rest of the vehicle usually drops into the ocean or burns up in the atmosphere, although some modern rocket boosters can be brought back and used again.

When you see a rocket launching, it starts by rising straight up from the ground. That makes sense, because space is straight over our heads. But if a rocket only goes up, when it gets to space and the engines are turned off, gravity will pull it straight back down again. That would be a pretty short trip.

So, rockets head upward at the beginning until they rise up above the thickest part of the atmosphere. At that point, they switch and arc overhead toward the horizon because they need to outrun gravity. In order to outrun gravity and use it to help you get around in space, you have to go fast. Very fast—thirty thousand kilometers per hour, to be exact. That's ten times faster than the speed of a rifle bullet.

So a rocket must go from a standing start to faster than a speeding bullet in only eight minutes. That's not a lot of time, and for anyone riding the rocket, it's quite an experience. You lie on your back facing the open sky as a huge rumble begins far below. At liftoff, you are pushed into your seat as you shoot straight up, accelerating faster and faster every second on a ride that is like no other. Everything shakes as your body gains weight from the increasing speed. Some astronauts say it's like an elephant is standing on your chest.

Once we launch a rocket into the air, we need to get it to space and make it stay up there. The path that rockets make through the air is a very special curve. It's called a ballistic curve, and everything that falls freely through the air without wings—from basketballs and baseballs, to bullets and rockets—follows the same kind of arcing curve.

When you throw a ball, it travels forward, but at the same time, gravity is always pulling it down toward the ground. The two forces acting together means that the ball follows a curve. Throw it harder, that curve flattens out a little and the ball travels a little farther. The harder you throw, the flatter its ballistic curve becomes.

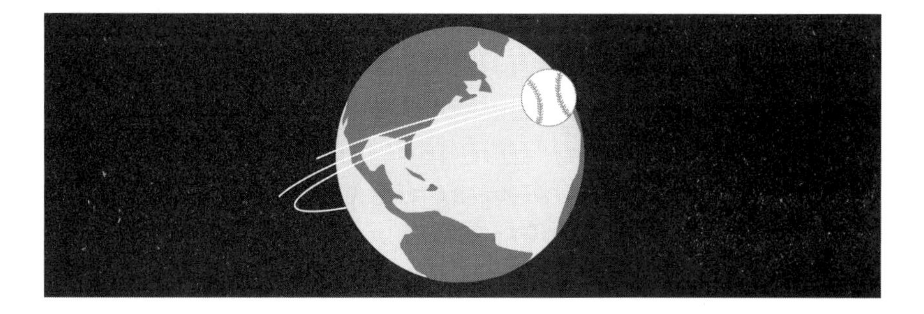

Now let's suppose you were a superhero and could throw the ball at super speed. A very hard throw would send the ball flying right over the horizon. And if you made a miracle throw that moved at thirty thousand kilometers per hour, something very special would happen: you'd be able to throw the ball right around the Earth.

At that speed, the curve that the ball follows is the same as the curve of the Earth itself. Gravity is still pulling the ball down, but it never hits the ground. Instead, it would fly all the way around the planet, eventually ending up right back where it started! An orbit, then, is just a ballistic curve that has wrapped itself in a circle that goes all the way around the Earth.

It would take about an hour and a half for the ball (moving at thirty thousand kilometers per hour) to make the trip around the world. If, after ninety minutes, you were still standing in the same spot wondering where the ball went, you could get hit in the back of the head!

It doesn't matter what you throw into space, if you can get it above the atmosphere and moving at the right speed of about thirty thousand kilometers per hour, that object will orbit around and around the Earth and not come down. In spaceflight, what goes up . . . stays up.

Throwing a ball into orbit is one thing. But what if you wanted to put yourself into orbit to meet someone else who was already up there?

Suppose you want to visit the International Space Station. The station is already in orbit, following a big circle around the Earth. At the same time, the Earth is turning beneath it. So you have to wait until your launchpad passes right under the orbit of the station, then take off at just the right moment so that you can catch up to the station along the same path. If you launch at the wrong time, you will miss it. That's why rockets have a "launch window": a short period of time when they have to get off the ground. If they miss the window, they have to wait for the next one to come along, which is often the next day.

It usually takes more than a day of circling the Earth to catch the space station. It's similar to runners in a relay race. As one person is speeding around the track, a second runner, waiting on the same track, starts out as the first one passes by, matching their speed and accepting the baton. In space, this game of orbital catch-up is called a rendezvous.

To go farther than Earth's orbit, you have to do the same thing, just on a bigger scale. Let's say you want to go to Mars. The Earth has a particular orbit around the sun, but Mars, being farther away from the sun, is following a different, bigger circle around our central star and traveling through space at a different speed than Earth—planets that are closer to the sun move faster than those that are farther away.

So, to travel between the two planets, you first have to wait until both of them are on the same side of the sun. Mars takes twice as long to circle the sun as we do, so sometimes the Earth gets ahead, like a speed skater on the inside track. You don't want to make the trip when we are on opposite sides of the sun because it would take more than a year to make the journey, which is far too long. If you make your trip when the planets are closest together—still fifty million kilometers apart—it takes only six or seven months.

To get to Mars, you'd have to fire your rockets even faster than if you're just orbiting Earth. But you couldn't just point your spaceship at your destination. Remember, Mars and Earth are both moving. That means you'd have to aim at the spot where Mars would be by the time you arrive. So you'd calculate the speed of Mars, the speed of the Earth, and the speed of your rocket to make it all work.

When you reach Mars, you'll be captured by the gravity of the planet. At that point, you'll either go into orbit around it or you might land on the surface. Hope you have a great time there. But don't forget to come home.

To do that, take the same path in reverse. Wait until the Earth comes close, and then spiral back around the sun, making your circle smaller and

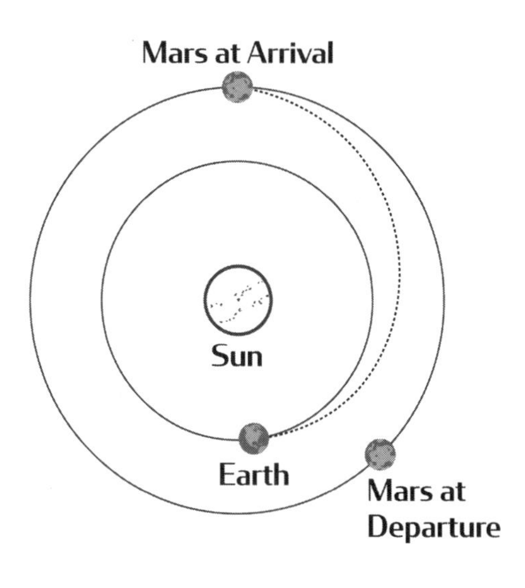

smaller until you meet your home planet. This is not as easy as it sounds because if you leave too early or too late, or if you go too fast or too slow, you can miss the Earth altogether. Make a mistake and you'll be lost in space!

Obviously, we have yet to send a person to Mars, but maybe one day we will, once we figure out how to do so safely. For now, only unmanned spacecraft are sent from one planet to another, and those spacecraft can take months or even years to reach their destinations. And because the solar system is so big and the spacecraft are traveling so far, it takes just a little error to miss the target. More than one spacecraft has missed its planet and ended up wandering around the sun as a piece of space junk!

Today's rockets are not very efficient. In many ways, they're like bullets shot from a gun—they get all their energy at the beginning of their flight and spend the rest of the journey coasting on their own. Sure, they get you to space, but they use up a lot of fuel, which makes them expensive to run. They also don't run very long. Those thirsty rocket engines gobble up all of their fuel in only a few minutes.

But it doesn't have to be that way. Engineers are working on other types of engines that burn more slowly and for a longer period of time. One such, called an ion engine, doesn't put out a lot of power, but it can run for more than a year without stopping. It's a little slow at the beginning, but after a year of pushing and pushing, it can actually end up going faster than the rockets we use now.

Our planet has been called *Spaceship Earth* because it's flying through space at more than one hundred thousand kilometers per hour as we orbit the sun. That's thirty kilometers every second! What if we could use the speed of the Earth around the sun as a slingshot to boost our spacecraft to other planets?

If we sent our spacecraft on a long orbit that includes the ship looping back to the Earth, the gravity of the Earth will pull the spacecraft ahead so the speed of the Earth through space is added to the speed of the spacecraft when it passes by, giving the ship the extra boost it needs to make it to another planet. It takes a lot longer to get to its destination, but it uses a whole lot less fuel.

This maneuver is called a gravity slingshot. The *Voyager* mission—which sent twin robot spacecraft to Jupiter and Saturn—was the longest journey to use this kind of energy transfer. One of the twins, *Voyager 2*, took advantage of the fact that all four of the largest planets in our solar system lined up on the same side of the sun at the same time. *Voyager 2* was first sent to Jupiter . . . which threw it to Saturn . . . then the gravity of Saturn tossed it to Uranus . . . and Uranus boosted it to Neptune . . . and Neptune kicked it right out of the solar system. It took *Voyager 2* twelve years to get to Neptune. That may seem like a long time, but without the gravity slingshots from all those planets, it would have taken more than twenty years to make the trip.

Did you know that there are two moons within the rings of Saturn that come so close to each other you could actually hop from one to the other? The moons are called co-orbitals because they both follow the same orbit around the planet.

Every now and then, the two moons approach on what appears to be a collision course and pass very close. As they do, their gravity starts pulling them together. The moon following behind is pulled ahead faster, thanks to a little gravity slingshot by its partner. At the same time, the moon that's ahead is dragged back and slows down. Basically, the two moons change positions. The lower moon is boosted to the higher orbit. The upper moon is slowed to the lower one. After this, they continue around Saturn until their next meeting, when they switch again.

These moons are tiny little ice balls only a few kilometers across. If you stood on one, you would weigh almost nothing. The smallest step would

ON THE DRAWING BOARD

Scientists are developing a plasma rocket. It uses extremely hot gases that rush out of the engine at a super-high speed, providing thrust. These engines can run for long periods of time—weeks or even months—always pushing spaceships faster and faster, up to speeds that might be fast enough to help us reach other stars. But plasma is extremely hot and can melt the engines. Scientists are working on ways to contain the fuel so the rockets can be used safely.

send you floating off the surface. So imagine standing on one moon as the other one approached. It would look like the sky is falling as the big ice ball passed overhead. If you jumped straight up at the moment the moon passed over, you could leap to it, landing on it like a fly on the ceiling. That would be your new home until the moon passed close again. Moon hopping—a sport of the future!

SPACE PLACES

When astronauts fly in space, they get a lot of attention, but robots can also go where no human has ever gone before. If you want to fly a robot to another planet, one place you have to do it from is the Jet Propulsion Laboratory in California. The JPL is mission control for robots that have been sent to every planet in the solar system. Here, scientists and engineers gather to design, build, and fly the robotic spacecraft that visit other worlds. With colorful names such as *Mariner*, *Voyager*, *Opportunity*, *Spirit*, and *Curiosity*, these mechanical explorers have ventured farther and been to more places than any human. (The farthest people have been in space is to the moon, which is in our backyard compared to the other planets.)

When a robot visits a planet for the first time, it simply flies past the planet once, taking as many pictures as possible on the way. Then, other robots follow that go into orbit around the planet, where they remain for years, mapping the surface, watching the seasons change and capturing images of the planet's moons. Finally, there are the landers that touch down on the surface and the rovers that drive around to show us what we will see when people finally do make it into deep space.

Every flight involves figuring out where the other planets are in their orbits compared to Earth, when the best time is to leave Earth so the robot takes the shortest time to make the trip, and when to arrive so they can hit an exact landing spot on the ground of the other planet. That's a lot of planning! But it's worth it for the views of alien worlds that our electronic partners provide.

YOU TRY IT!

Blast Off

You can build a simple model rocket using only a few ingredients and materials. Here are three different versions.

WHAT YOU NEED

1. A long, thin balloon
2. Baking soda
3. Vinegar
4. An empty plastic bottle with its cap (two-liter soda bottle is best)
5. A pair of scissors or a knife
6. A skateboard or rollerblades
7. A heavy rock

WHAT TO DO

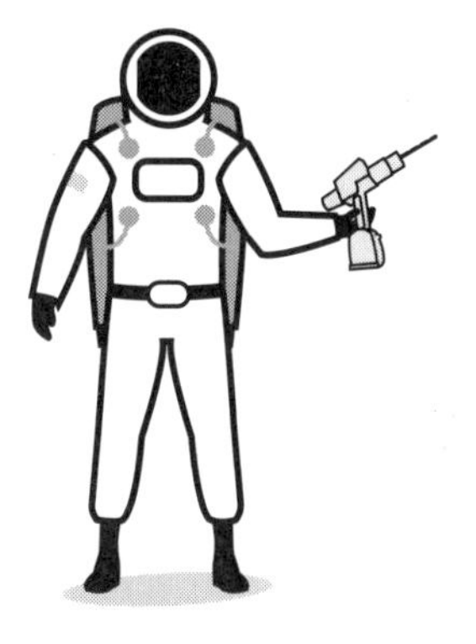

1. Go outside. Blow up the balloon. It's about to become your rocket. Point the neck to the ground and let it go. The air rushing out is the action; balloon flight is the reaction. You have probably done this at a party, but it is actually rocket propulsion!
2. The second model rocket is made from a plastic bottle. Take the bottle, baking soda, and vinegar outside.
3. Using the end of a pair of scissors or a knife, carefully cut a small hole in the center of the plastic cap. Make sure the hole goes through the soft liner on the inside of the cap.

4. Pour baking soda into the bottle so the bottom is covered about two centimeters deep.
5. Pour a generous amount of vinegar into the bottle, then quickly screw the cap on.
6. Place the bottle upside down on the ground, leaned against something so the bottom of the bottle is pointing up. Stand back and admire.
7. Experiment with different amounts of baking soda and vinegar, as well as the size of the hole, to get the maximum thrust.
8. Your third rocket model is made with your own body! Stand on a skateboard or wear rollerblades and pick up a large rock, as heavy as you can lift. With your wheels stopped, throw the rock straight out in front of you as hard as you can. You should find yourself rolling backward in the opposite direction. Throwing the rock one way was the action, like a rocket exhaust; the motion of your body in the opposite direction was the reaction. A rock rocket!

Reaction
Action

20

How Do You Walk in Space?

Taking a walk in space is not as simple as opening a door and stepping outside. In fact, you don't step out at all; you float out because you are weightless. There's no air to breathe in space, and radiation from the sun would roast you if you went out there in your normal clothes. So before you open the hatch on a spaceship, you need to learn how to put on a bulky space suit, operate special tools while wearing stiff gloves, handle large objects that would be too heavy to pick up on Earth, and do construction work while floating weightless four hundred kilometers above the Earth with nothing below you.

Sound complicated? It is. It takes years of extra training to walk in space. But the view you get of the Earth and the rest of the universe is worth it.

A space suit is more than a fancy outfit. It's a one-person spaceship that you wear. It provides you with air to breathe, protects you from harmful radiation, keeps you warm and comfortable, has a radio so you can talk to others, and it even comes with little thrusters so you can fly back to the space station if you drift away—and it has to do all that while being flexible enough for you to work in it.

Space suits evolved from the flight suits used by jet fighter pilots. The first ones were shiny and had a long hose, called an umbilical, that was attached to the spacecraft to provide the astronaut with oxygen. Later, the suits became bigger, with oxygen carried in packs on the back so that

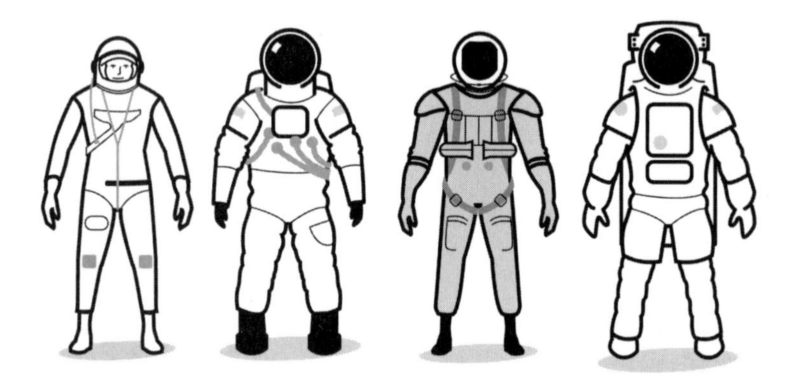

astronauts could move freely anywhere, and had many more layers to protect against micrometeorites—small pieces of dust that are always flying through space. You might not think a piece of dust would do much harm, but when it is moving more than thirty thousand kilometers per hour, it can go right through a suit. So underneath the white fabric of today's space suits are layers of aluminized plastic that act as a shield for the astronaut.

In the future, space suits will be more rugged and more flexible so they can withstand the rocky, dusty conditions of the moon and Mars. They might also fit skintight, like the wet suits that divers wear, providing even more flexibility for climbing cliffs or exploring caves. *Apollo* astronauts who landed on the moon found their suits quickly became covered in dark dust that clung to everything. They were concerned that, when they came back inside their lander, the dust would get into their equipment and cause problems. Fortunately, that didn't happen, but one concept for future suits is to always leave them outside. Each suit might have a hatch on the back with a door in it. That hatch would attach to the outside wall of a Mars habitat or the front of a lunar rover so astronauts could crawl in and out of the suit without bringing any dirt inside the vehicle.

Another thing you should know about outer space: it's cold out there! No, it's hot out there! No, it's both!

Space itself is cold. It can reach temperatures of minus 150 degrees Celsius in the shade. At the same time, the sun shines really brightly because there are no clouds or air to act as filters for the strong rays. In space, anything the sun shines on heats up to 120 degrees Celsius. But it only gets hot on the sunny side. Without air to spread the heat around, like it does for our bodies on Earth, an astronaut's shady side remains super cold even as the sunny side is super hot. That means there can be almost three hundred degrees of temperature difference between the two sides of an astronaut in space. That's one reason space suits are white. White reflects sunlight away, so it is not absorbed and turned into heat as much.

To make sure astronauts don't freeze and fry at the same time, they're surrounded in water. An undergarment that looks like long underwear has little plastic tubes woven into the fabric with water running through them to keep the astronaut's body the same temperature on all sides. The temperature of the water can be adjusted up and down. When a spaceship is on the night side of the Earth, it's in the shadow of the planet where it's very cold—time to turn up the water heat in the astronaut suit! But wait: the ship is back to the sunny side again—time to cool that suit down. During a space walk, the temperature in the space suit is constantly being adjusted by the astronaut.

While we've had many great successes with space walkers over the years, it remains a very perilous activity. It's safe to say that walking in space is no walk in the park.

On March 18, 1965, Russian cosmonaut Alexei Leonov crawled through the hatch of his *Voskhod* space capsule and became the world's first space walker. Floating in the emptiness of space, more than four hundred kilometers above the Earth, he exclaimed, "I feel great!"

He was outside for only twelve minutes, but they were almost the last twelve minutes of his life. Soon, he felt his hands leaving his gloves, and his feet rising out of his space boots. His suit became stiff as it inflated like a giant balloon. His suit wasn't supposed to do that, but no one had ever worn a suit in a place where there was no air on the outside. The air inside the suit was pushing outward, and the fabric was too loose, so the suit was slowly getting bigger. Leonov found it hard to move his arms and difficult to pull on the long cord that attached him to the spacecraft. As he struggled to get back in, his body temperature rose dangerously high until he was covered in sweat.

When he did manage to get to the capsule and began entering the hatch, he found the space suit was too big to fit inside. The suit wouldn't budge. Leonov was exhausted, but he couldn't remain outside. You can't return to Earth hanging on to the outside of a spacecraft.

Finally, in a desperate move to get through the tight opening, he let some air out of his space suit so that it would become softer. While losing precious oxygen and on the verge of passing out, he managed to squeeze back inside his space capsule and seal the hatch. It was a narrow escape!

Today's space suits are built stronger. They don't inflate like balloons, but they still become stiff when out in the vacuum of space. Astronauts must learn how to work with the suit so they don't become exhausted. It looks easy to be floating around in zero gravity, but make no mistake: space walkers are working hard!

Before you can walk in space, you have to learn to walk underwater. It's the only other way to simulate the weightlessness in space for a time period longer than thirty seconds, which is all an astronaut in a zero-g airplane gets.

The world's largest swimming pool is at the Johnson Space Center in Houston, Texas. The pool has to be big because inside it are full-size models of the space station. Astronauts go underwater and practice the work they'll be doing in space. Wearing the same type of space suit used in orbit, space walkers are lowered into the water, where divers add weights to the suit until the astronaut neither floats nor sinks. Once an astronaut is neutrally buoyant, they float in the water somewhere in the middle—not at the surface or the bottom of the pool—similar to the way they would float in space.

Divers guide the astronauts to the space station and hand them the tools they will be using in outer space so they can practice using them while weightless. On Earth, if you need to loosen a bolt, you put a wrench on it and turn it, usually with quite a bit of force. But if you're floating weightlessly, your body force isn't much help. The wrench and the bolt remain still while you spin around. That's not very useful.

Space walkers have to learn what parts of the space station they can hang on to or where to brace their feet while working with tools, which in some cases means turning upside down. Luckily, that's easier to do in space because there is no up or down.

Even though it is called space walking, it's not done with the feet. Astronauts actually walk on their fingertips in space. In zero gravity, your feet don't stick to anything, so your hands and arms do most of the work of moving you around. There are restraints or hoops that you can loop your feet into to stay in one spot, but that is space standing, not space walking.

The International Space Station is larger than a football field, and believe it or not, it's possible to get lost while floating around outside. A tool that helps astronauts find their way around is a virtual reality room at Johnson Space Center that reproduces the entire station in 3-D.

I was lucky enough to visit and try it out. Wearing a VR headset, a chest pad that monitors body position, and wired gloves, I sat in a swivel chair with Canadian astronaut Jeremy Hansen in a chair behind me.

"We are putting you on the very end of Canadarm 2," the operator said.

As the system turned on, I saw nothing but blackness, then two white space-gloved hands appeared in front of me as I reached out. I watched what looked like my gloved fingers open and close at exactly the same rate as my own hands in the room. I made the gloves wave around like a baby reaching up from a crib.

"I don't see the space station," I said into my headset.

"Try looking down."

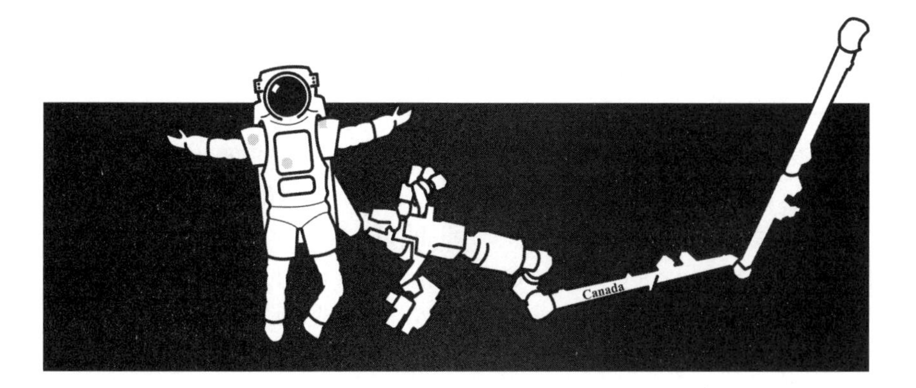

I tilted my head down toward my feet, and there was the entire sprawling complex spread out below me. Silver cylinders the size of buses hooked together end to end, and long, thin solar panels reached out the sides, golden in the sunlight. I recognized the American module straight below me, the Russian module off to the left, and the Japanese and European sections off to the right. I was virtually standing on the tip of a robotic arm made in Canada! This truly was amazing!

But there was something missing. Everything around the station was black. Where was the Earth? I'd heard so much about how beautiful the view of our planet was from the station, but it was nowhere to be seen in this VR simulation. Then I remembered that I was in a fully three-dimensional space.

Leaning back and tilting my head up as far as I could, the vast expanse of the Earth completely filled my view. And blue was all I saw. We were flying over an ocean. A thin, wavy brown line passed by. It must have been a beach on some island, but I had no idea where I was. I did realize that I was actually positioned upside down under the space station with my head pointed toward the Earth. But then again, there is no up or down in space, so it doesn't really matter.

"Okay, we're going to put you on the structure now," said the operator, and with a click of a mouse, I was face-to-face with a wall of gray metal. I looked around and saw nothing but beams and girders in all directions. I had no idea where I was on the station because I was now stuck to the side of it.

"Now get ready. The sun is setting. We're going around to the night side."

Right. I knew that the space station goes around the Earth every hour and a half, so astronauts experience sixteen sunrises and sunsets every day—one every forty-five minutes. The darkness fell as quickly as the lights dimming in a theater. Everything around me disappeared into blackness except a circle of light directly in front of me that came from my space suit's headlights. It illuminated a small part of the giant structure, and as I turned my body, I could see only what was in that white circle of light.

Space walkers must work in the dark as well as daylight, I suddenly realized. They have to find their way around the outside of the huge station while floating freely in a bulky space suit using their hands to maneuver, then work with tools and move huge pieces of equipment that would be impossible to lift on Earth. They work in the most dangerous place imaginable on the largest objects ever sent into space. It takes strength, concentration, and knowledge to pull it off.

I didn't get a glimpse of the Earth at night because I was too busy trying to find my way around the maze of the space station structure in the dark. But I did think about the blackness that surrounded me and how far into space it goes. I tried to imagine the feeling of floating alone out in the void with nothing but the fabric of the suit and a thin plastic faceplate between me and the rest of the universe. It would be a humbling experience.

When the simulation was over and I removed the headgear, I felt like I had been on a remarkable journey. The room seemed small and confining. Gone was the feeling of being in a vast open space. I was an Earthling bound to my planet once again.

Working in space—there really is no job like it.

YOU TRY IT!

Space Construction

Space walking is construction work done in space. Astronauts leave their spacecraft to do tasks and research, and that means using tools while wearing a bulky space suit. It's not as easy as it looks. Why not give it a try?

WHAT YOU NEED

1. An assortment of tools—screwdrivers, pliers, hammer, pen and paper, paper clips, a knife, a fork, and a spoon (or whatever implements you can find)
2. Assorted screws, nails, nuts, and bolts
3. A scrap piece of wood
4. A bowl
5. A box of cereal
6. A carton or jug of milk
7. Thick winter gloves or hockey gloves

WHAT TO DO

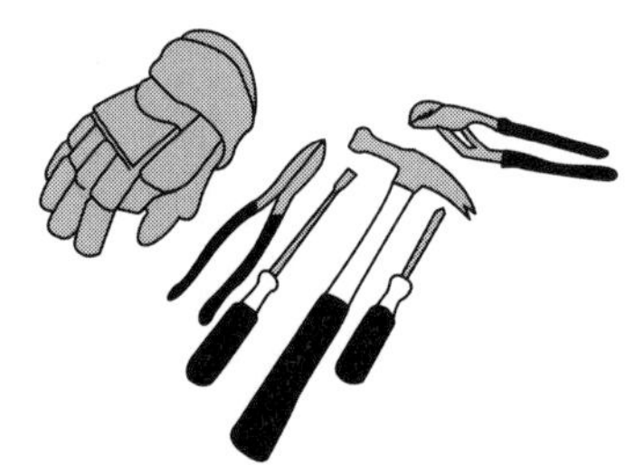

1. Put on the gloves.
2. Grab the screwdriver and one screw.
3. While wearing the gloves, try driving the screw into the wood with the screw-driver.
4. Once you've accomplished that, grab the pen. Keeping your gloves on, write a note on the paper, trying as best

you can to keep your handwriting neat. Can you read what you wrote?

5. Next, open the cereal box and pour yourself a bowlful. Open the carton of milk and pour some on your cereal without spilling. Now try to eat it with the gloves on.
6. Congratulations! You are now a tool-using space walker.

How did you do? I hope there's no spilled milk! Now imagine what it's like for astronauts working with tools in zero gravity. And remember: you can't drop anything in space. If you do, it will float away and you'll never see it again!

21

When Will We Go Back to the Moon?

It has been more than fifty years since the first human beings landed on the moon. On July 16, 1969, the mighty Saturn V five-moon rocket—the largest machine to ever fly—slowly lifted off the launchpad, riding atop a pillar of fire. In a small capsule on the tip of the rocket were Neil Armstrong, Buzz Aldrin, and Michael Collins.

Four days later, on July 20, Armstrong and Aldrin guided their lander, *Eagle*, onto the surface of the moon, and the famous "one small step" was taken by Armstrong. He and Aldrin became the first humans to walk on another world. For a moment, the world was united as people everywhere shared in humanity's greatest adventure.

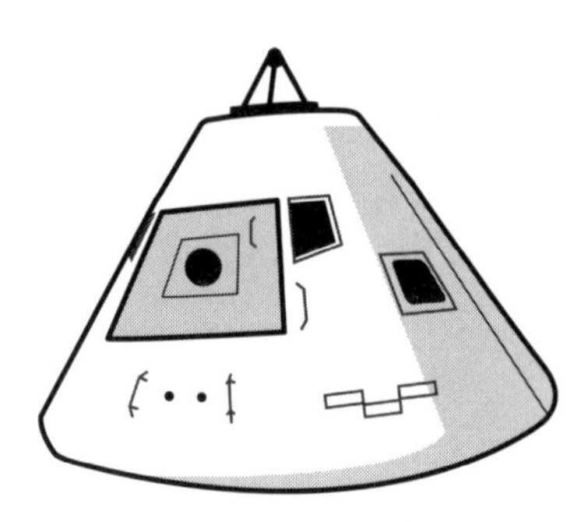

There were five more lunar landings after Armstrong and Aldrin made the first boot prints in alien soil. In total, twelve men have actually walked on the surface of the moon. Some of them have even driven an electric car among the craters. The last footsteps were from Gene Cernan in 1972. No one has been back to the moon since.

American astronauts are slated to return to the moon sometime in the 2020s, first in a new moon space station, then down to the surface.

The type of rocket that went to the moon—the mighty Saturn V—was twice as high as Niagara Falls. A new rocket called the SLS, or Space Launch System, is planned to be even taller. And other rockets such as the Falcon Heavy and Starship, both from SpaceX, are almost as big.

The small capsules that carried astronauts to the moon were called the Command Modules. They held three people. A new Starliner capsule has the same shape but is much larger and carries a crew of four to six. And another, called Dragon 2, holds seven.

You might be wondering why a new spacecraft looks the same as one that was used half a century ago. Why not something with wings?

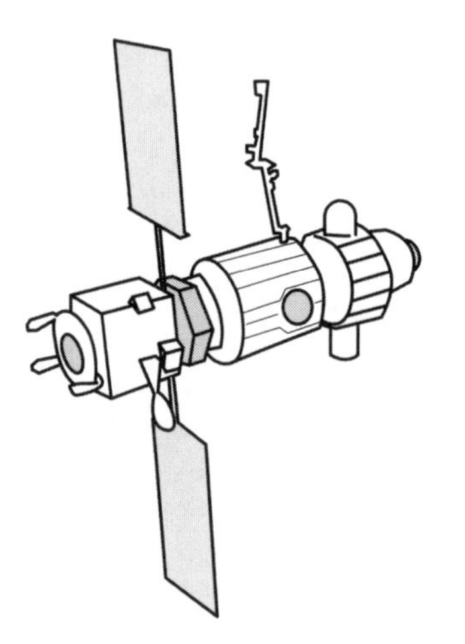

The short answer is that it worked before, so it should work again. Since it takes a lot of energy to get to the moon, something small without wings is easier to fly. Besides, you'll recall that wings are only used to land back on Earth. There's no air on the moon, so why bring wings?

Here's how it will work. The first trip back to the moon will not actually be to land on it but to build a space station, Deep Space Gateway, that will orbit around it.

Like the International Space Station but smaller, Deep Space Gateway will be built out of linked modules made by many countries. It will be a home for astronauts to study the moon from above before anyone goes to live there.

Once Deep Space Gateway is finished, the next missions will be able to focus on returning to the surface of the moon. The world's largest rockets—the SLS, the Falcon Heavy, or the Starship—will carry a crew and capsule into orbit around the Earth. Once in space, the astronauts will take off on a three-day journey to the Deep Space Gateway. When they arrive, the crew will rendezvous with the other astronauts who work at the lunar-orbiting station and pick up their lunar lander. Some of the crew will enter the lander while the rest remain in the station.

Landing on the moon will be its own exciting challenge. Since there will be no air, wings or parachutes won't be any use. Instead, as the lunar lander descends, it will use its rocket engine to slow itself and drift down slowly, like a helicopter. Landers have unusual shapes because they don't have to be aerodynamic. They don't need nice, smooth bodies like airplanes. The lander will kick up lots of dust, but hopefully it will manage a clean landing. Once it does, astronauts will finally step out and onto the surface of the moon once again.

Back on planet Earth, we will rejoice. We'll look up into the sky in wonder. We'll celebrate. But there will be one big difference between this lunar mission and the ones in the old days: this crew will be on the moon to stay. They will be there with a habitat, living quarters, and laboratories to live for months at a time so that they can conduct scientific experiments, collect samples, and test equipment that will eventually be used to explore Mars.

The lunar habitat will be made of similar modules to the Deep Space Gateway. They'll be hooked together but will lie on the ground rather than

float in space. The habitat might even be inflatable, ready to be blown up once the astronauts get there.

One thing that's nice about living on the moon as opposed to Mars is that astronauts will see the Earth out of their habitat window. The moon also offers a wonderful viewing platform for the stars because there is no atmosphere or clouds, so the view of the universe is incredibly clear.

It's also a chance to explore more of the moon itself. There's one area of the moon that may be the most unusual and interesting of all: the South Pole. There's a crater at the bottom of the moon where the sun never rises above the rim. Remember, the sun shines mostly around the equator of the moon, just like it does on Earth, so at the South Pole, the sun is very low in the sky. That means the floor of this crater is always in shadow. Shadows on the moon are extremely cold—cold enough for ice to exist. On the moon, ice would be as valuable as gold because it would provide drinking water and could be used to make rocket fuel for an eventual voyage home.

But ice on the moon, again like gold on Earth, is not easy to get to. The ice exists at the cold and dark bottom of the deep moon crater, which is hard for machinery and astronauts to reach. On the other hand, the edge of

the crater—the top of the rim—is in sunlight all the time. We see the same effect on Earth during the summer in the high Arctic, where the midnight sun stays up for months at a time.

One idea is to place solar panels at the top of the crater rim to produce electricity from the sun, build a colony there, and send robots to mine the ice down in the crater floor of the moon. You might have heard of water as H_2O. That means it is made of hydrogen and oxygen. If water is taken apart, using electricity drawn from those solar panels, then the oxygen can be used for breathing and the hydrogen can be burned as rocket fuel.

Living on the moon means more than just work all the time. There has to be some playtime as well. So far, only one person has played a sport on another world. In 1971, Alan Shepard, commander of *Apollo 14*, took two golf balls to the moon. Using a makeshift club, he fired off two shots. The first he flubbed, but the second supposedly went out of sight, thanks to the low lunar gravity. The problem with lunar golf is there's no turf. The moon is one giant sand trap. And with so many craters, how do you know which one is the hole?

Other sports could be just as fun. The moon is smaller than the Earth, so gravity is not as strong there. If you lived on the moon, you'd weigh one-sixth your weight on Earth. Imagine the trick shots you could make in basketball, or how high you could leap in the high jump. What if we melted some of that moon ice and made it into a skating rink for figure skating? Quad jumps? Heck, you could probably spin ten times before coming back down. In fact, you'd be jumping so far, they'd have to make the ice surface a lot bigger. The first Olympics on the moon would break all records!

On the other hand, another group of space scientists thinks it would be a waste of time going back to the moon because we have already been there. They think we should skip the moon and go straight to Mars.

What do *you* think?

YOU TRY IT!

Moon Habitat

WHAT YOU NEED

1. Chairs, tables, blankets, and sheets
2. Two large pieces of cardboard
3. Packaged food supplies for twenty-four hours
4. Pillows and blankets (or a sleeping bag)
5. A computer loaded with games and information about the moon

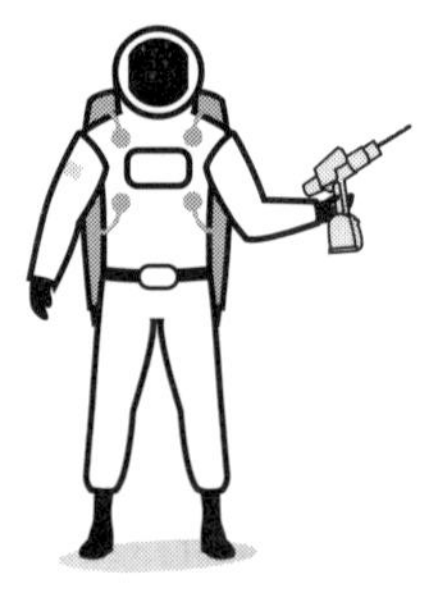

WHAT TO DO

1. Using the furniture and blankets, build a habitat as large as possible with three sections that are connected together: section one is the smallest and is for sleeping only, section two is the food storage, and section three is the gathering room.
2. Place the cardboard sheets between each of the sections of the habitat to simulate sliding doors.
3. Try to remain in the habitat without coming out for twenty-four hours. (Bathroom breaks are allowed, but you have to return immediately without getting any supplies from the house.)
4. During your stay, imagine you are really on the moon. Find out as much as you can about your new home. Look at maps and find places you would like to visit, such as the

dark craters at its South Pole that contain ice. Plan expeditions and talk about what kind of equipment you would need.

Building a habitat on the moon will take a lot of planning because the crew will need to bring everything with them from Earth. Habitats will also be small at the beginning—about the size of a small house—which means that people will have to live together in close quarters. There is also no internet on the moon (yet), so entertainment will have to be brought along in the form of games and predownloaded programs.

Thankfully, the moon is close enough to the Earth that you will be able to make calls to your friends back home. On Mars, that will not be an option because the red planet is so far away. It can take up to twenty minutes for a signal to reach Earth and another twenty minutes for a return, so conversations from Mars will have to be one-way messages or videos.

22

When Can I Take a Vacation in Space?

Everyone who has flown in space has had fun up there. Floating around weightless like Peter Pan, playing with food, performing crazy stunts never possible on Earth. Who wouldn't want to join in on the fun?

If you want to fly in space today, there are three ways to get there. You can climb aboard an American rocket. You can ride a Russian Soyuz rocket. Or you can take a flight on the Chinese Shenzhou rocket. These are the only ways to reach space at the moment, and to fly in any of them you have to train for years.

But all that is about to change. Soon, you'll be able to take a spaceflight just for the fun of it, and you won't have to be an astronaut to do it.

Companies around the world are building a new generation of spaceships—small, cheap rockets designed to take tourists on the vacation of a lifetime. Virgin Galactic SpaceShipTwo is a combination of a rocket and an airplane that seats six passengers and two pilots. It's carried high in the air by a large airplane called WhiteKnightTwo. The spaceship is released from the plane, fires its rocket engines, and takes the passengers on a thrilling high-speed ride almost straight up above the Earth's atmosphere to the edge of space. When the rocket engine shuts down, the passengers unbuckle their seat belts and float around weightless for a few minutes, enjoying the view of the Earth and the blackness of space.

After the space plane reaches its maximum height, it begins to fall back to Earth. Everyone returns to their seats and hangs on for an exciting dive back down into the atmosphere before they glide back to the same runway the space plane took off from.

Another spaceship, called New Shepard, uses a capsule with very large windows that sits on top of a rocket. The ship blasts off straight up from the ground and goes just above the atmosphere, where the capsule separates and the people inside experience a few minutes of weightlessness. Then both the capsule and rocket come straight back down separately, with the rocket booster using its engines to land and the capsule floating down by parachutes.

Both of these spaceships are designed to take tourists into space and are operated by private companies. That means anyone can buy a ticket to space.

If you want to spend much more time up in space, you can even buy a ticket up to the International Space Station. You would travel to Moscow, to train with the Russian space agency, then fly to space on a Soyuz rocket alongside professional cosmonauts. After a week on the space station, you would come back with another crew who had been living in space and are on their way home. Several space tourists have already made the trip. One thing they had in common: they were all very rich. That's because a ticket to the International Space Station costs about $50 million! Tickets for the shorter tourist flights will be much cheaper: only $250,000.

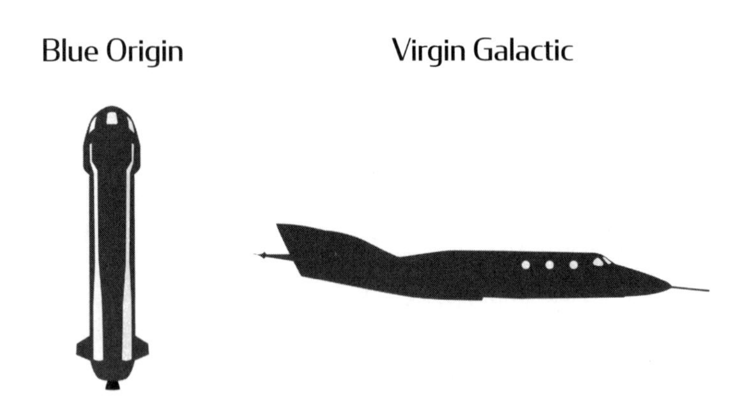

The cost of flying in space might seem expensive right now, but when airlines first started many years ago, only rich people could afford to fly. After a while, as more people took to the air, the price of a plane ticket came down. Today, many people can afford to fly on an airplane. Maybe the same thing will happen with spaceflight. We'll let the rich people get it started, then the rest of us can fly later.

In the future, it will probably be possible to spend a week at an actual hotel in space. It won't be very big, but it will be made just for tourists, so you'll have your own room and a window to look down on the Earth. Eventually, more modules will be added, which means there will be more room and more things to do up there.

But these small space hotels are just the beginning. Imagine really large structures shaped like giant Ferris wheels that slowly turn. In your room along the rim, there would be artificial gravity created by the spin, so you'll be able to walk around like you do on Earth, except that you'll be in space.

While you're there, if you want to experience weightlessness, just go to the center of the wheel, where you would weigh nothing at all. That would be the best place to put the space swimming pool. Can you imagine what the pool would be like in a space hotel? Because of the weightlessness, the water in a regular pool would float all over the room! But if the pool were in the shape of a big drum that slowly spins, the water would stick to the inside of the walls. Think of the fun you could have swimming all the way around the walls and ceiling and doing slow-motion dives!

Vacations in space are just in their infancy. For now, there's no space

hotel or weightless swimming pools, and there are no trips to the moon . . . but even that is coming soon. The one giant leap into space you may be able to take in the future—and it is really out of this world—is, in fact, a moon trip. You won't land or get to walk on the moon, but you'll get to go there, fly around it, and return home again. The trip would take about a week. No tourist has made the trip yet, and that could be because of the cost. A ticket to the moon costs only . . . one hundred million dollars!

While we wait for commercial spaceships to be built to take us to the moon and beyond, there's a way you can feel like you're flying in space and experience the weightlessness astronauts do by flying in a very special airplane made by a company called Zero G. It uses a big airplane affectionately known as the "Vomit Comet." It's called that because some people get sick riding in it. I was fortunate enough to fly in this plane, and happily, I did not get sick!

The airplane is an airliner with most of the seats removed and all the windows covered up. You sit in seats that are at the back of the plane, and as it takes off, you can't tell what the plane is doing because you can't see outside. You're inside a big hollow tube. All you can hear are the jet engines.

When the plane reaches a special zone over the ocean, everyone gets out of their seats and lies on their backs on the padded floor. The roar of the engines gets louder as the pilots point the plane upward at a steep angle.

Then the plane is put through a special arc called a ballistic arc, which is the same type of curve that a baseball, football, snowball—anything that is thrown through the air—follows. And remember: when anything is falling through the air, it's weightless until it hits the ground.

As the plane comes over the top of the curve, everyone inside rises off the floor and floats weightless inside. It's as though someone simply turned gravity off and your body no longer stays down. Everyone in the plane I was on immediately began to laugh hysterically as we bumped into one another and slowly bounced off the walls. I had an urge to wave my arms and kick my feet as though swimming through water, but that doesn't work in the air. In fact, the more you wave and kick, the more you tumble out of control, with a good chance of kicking someone else in the head.

It takes very little to set your body in motion. Just the slightest touch on a wall or the floor sends you floating in the opposite direction, and once you start moving through the air, it's really hard to stop yourself until you hit something. Soon, you learn to slow down and take your time, which is why astronauts seem to move in slow motion. They have to be gentle or they will spin out of control.

After thirty seconds of floating, the airplane ends up pointing steeply down toward the ground. Instructors in the plane with you yell out, "Feet down!" That means point your feet toward the floor. As the pilots pull the airplane out of the dive, you float gently back down to the floor, and as it continues through the bottom of the curve, heading back up for the next maneuver, gravity comes back, getting stronger and stronger until you weigh twice your normal weight from the force of the plane climbing. That's what it would feel like to be on a planet much bigger than the Earth, where the gravity is stronger.

The pilots also gave us the feeling of walking on a planet smaller than Earth, too. One g is the gravity of Earth, and right now, you're feeling one g as you sit in your chair. But Mars is smaller than the Earth, so the gravity is less there.

On one of the ballistic hops, the pilots brought our weight down to one-third of a g, which is the gravity of Mars. As the airplane simulated Mars, I found my walking style changed. Each step lifted me off the floor just a bit, so I could bounce along on my toes. I tried to imagine people on Mars all walking with a bounce in their step!

Next, the pilots lowered our weight even further to one-sixth of a g, which is the gravity of the moon. You've probably seen movies of the *Apollo* astronauts bouncing around in the low gravity of the moon, but they were wearing bulky space suits. In ordinary clothing, you feel really light in moon gravity. When I jumped, I floated back down slowly, like a leaf falling from a tree. That was a surprise. In a moon habitat, if you wanted to go up to the second floor, you wouldn't need to use stairs. You could easily jump up and float to the upper floor without hurting yourself.

In moon gravity, I was able to do a standing backflip in slow motion. When I tried to flip forward, I pushed off too hard. My back hit the ceiling of the plane and I seemed to stick up there for a second. Then I started to slowly fall back down headfirst. Luckily, I had time to put my arms out, touch the floor while upside down, and push myself slowly back onto my feet. It was amazing. What did I learn? Gymnastics on the moon will be a breeze!

It's quite a ride, changing from super heavy to no weight at all and back again fifteen times in a row. Not bad for a single plane trip! During each period of weightlessness, I got better at controlling my body, until I was able to do multiple tumbles end over end while floating in the middle of the space. It was a magical experience, floating freely like Peter Pan and being able to fly in any direction like Superman. No wonder astronauts seem to be having so much fun in space!

YOU TRY IT!

High-Flying Hijinks

You can experience a little bit of weightlessness every time you go over a bump on a bicycle or get air on a skateboard. If your wheels leave the ground and you're in the air, you're as weightless as an astronaut. Stunt cyclists are just like astronauts when they perform their high-flying leaps. They have no weight when they're falling, which is why they can do crazy maneuvers over the handlebars that would be almost impossible on the ground.

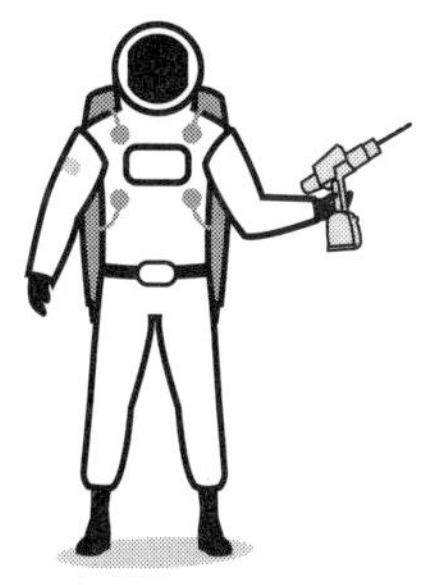

WHAT YOU NEED

1. A bicycle, skateboard, or motocross bike
2. Some courage and speed

WHAT TO DO

1. Get going fast and go over a ramp or bump that leaves you airborne.
2. For that brief moment that you're in the air, you're feeling the effects of zero g!

You can even feel a little bit of weightlessness in a car when traveling over hilly roads. Every time you feel your stomach lifting up, there goes gravity. So, as you can see, the sensations of spaceflight are not that far away!

23

How Do You Drive a Robot on Mars?

Robots are the true space explorers. They've been to more worlds more often than humans ever have. They've even gone where no human being ever will go—into the stormy clouds of mighty Jupiter, through the rings of Saturn, and out to the cold, dark reaches of Pluto. Robots went to the moon before people did to check it out and make sure that it was safe to land there.

But building robots and operating them on worlds that are millions of kilometers away from Earth is not easy. If something goes wrong with the robot, you can't run up and fix it. The robots we send into space have to be strong, capable, and somewhat smart.

So what is a robot anyway? The word "robot" goes back hundreds of

years to the Czech word *robota*, which means forced labor or drudgery. At first, people thought robots would work like human slaves and that they would look and act like we do. But robots of that variety are made mostly for entertainment—they're useful in movies, not on other planets.

When it comes to working in space, robots come in many different forms, depending on what they're meant to do. Some have solar wings and antennae sticking out in all directions. That variety is not built to land on planets but to fly past or orbit them and examine the planet from above.

Others have wheels so they can make landings. Rovers—such as *Spirit* and *Opportunity*, which both landed on Mars in January 2004, and *Curiosity*, which touched down there in 2012—have wheels to drive around and moveable heads with camera eyes so they can see where they're going. They also have an arm to reach out and sample rocks. The robots' eyes are up high—about the same height above the ground as human eyes—so the pictures they take show what the landscape would look like if you were standing on Mars. Rovers are the most complicated and talented robots that are currently sent into space.

Driving a car on Mars is not the same as driving a radio-controlled car on Earth. Mars is so far away that it takes time for a controller's radio signal to travel across space and reach the robot. The travel time between controller to rover is up to twenty minutes. If the controller sees the rover heading into danger and tells it to stop, the robot won't receive the message until it's too late. For this reason, robots need independent brains. They need some intelligence of their own to recognize danger and avoid it.

Some robots have all their actions programmed into their computers ahead of time so that when they arrive in space, the program runs by itself. The robot fires its engines and turns on its cameras and other instruments at exactly the right time. This takes a lot of planning ahead because if the timing isn't right, the robot might do the wrong thing at the wrong time and not even know it.

Rovers that drive around on other planets have such intelligent computer brains that some have compared them to obedient pets. They're given instructions on where to go and what to do, then left on their own to do those things. It's like a dog fetching a stick. The dog knows how to run after the stick, pick it up, and bring it back. All you have to do is throw the stick. The dog does the rest. Rovers on Mars operate the same way.

Scientists on Earth examine pictures the rover takes of the land around itself and decide where they would like it to go. If there's an interesting rock nearby, the controller will send signals to the rover to drive toward it. But while the rover is driving, it's also making sure it doesn't run into anything along the way. If it comes to a boulder that is too large to drive over, it will either decide to drive around the obstacle or stop and phone home to ask what to do next.

The rover uses cameras, lasers, and other sensors to look at its environment as it drives along, making a map of the land and choosing the best route. This takes a lot of computer power, which is why rovers move very slowly, about the speed you would if you were crawling along a bumpy floor on all fours.

But rovers move a lot more quickly when they're landing. Scientists call it the seven minutes of terror. That's how long it takes a robot to go from floating in space all the way to the ground on Mars. During that short time, the rover comes screaming into the top of the Martian atmosphere at more than twenty thousand kilometers per hour and has to perform a series of maneuvers involving a heat shield, parachutes, and air bags to come to a dead stop on the ground. If anything goes wrong on the way down, the mission is over. And during this dangerous descent, the rover is entirely on its own.

I was fortunate enough to be among a large group of reporters and special guests at the Jet Propulsion Laboratory in California, which was mission control for the landing of *Spirit* on Mars on January 4, 2004. Excitement filled the air because we were about to witness a landing on another world, where we would be the first to see an alien landscape that human eyes had never seen before. But first, we had to get the rover there.

The spacecraft, which looked a lot like a flying saucer out of a science fiction movie, held the lander *Spirit* inside it. The capsule had been launched from Earth about seven months earlier and spent all that time coasting through the cold vacuum of space between Earth and Mars. Now it was time for the hardest part.

More than half of the robots sent to Mars have failed. They either missed the planet, crashed onto the surface, or simply lost contact along the way. Landing on another world is not easy.

In a separate room, mission controllers were watching their computer monitors carefully. Everyone was hoping the mission would go according to plan. If something went wrong, the signal delay between Earth and Mars meant the controllers could not do anything about it. The rover was on its own—all the scientists and the rest of us could do was watch.

A mission controller announced, "Contact with the atmosphere."

A cheer went up in the room. The spacecraft had reached Mars. But the tension continued to rise as the rover, tucked into its protective capsule, was surrounded by hot plasma as the Martian air burned around it.

The round capsule functioned as a heat shield. It was designed to protect its precious cargo from temperatures that, because of the friction with the air, could reach more than two thousand degrees Celsius. To do that, the saucer had to hit the atmosphere at just the right angle. If it dove in too steep, it would burn up and be destroyed. If it came in at too shallow an angle, it would skip off the top of the atmosphere like a stone off water and bounce back into space. Thankfully, *Spirit* was right on target.

A fireball surrounded the capsule, turning the spacecraft into a bright meteor streaking through the Martian sky. That air friction also slammed on the brakes, slowing the spacecraft down and putting pressure on the lander. It was the same feeling as when you're forced against the seat belt of a car after a driver hits the brakes, except the rover felt many times the force of gravity.

After surviving the fireball stage, the capsule was still traveling more than two thousand kilometers per hour. A gigantic parachute popped open, slowing it down further. Even with that, though, the capsule was still eleven kilometers above the ground and moving faster than the speed of sound.

When the capsule was eight kilometers above the ground, the heat shield dropped off because it was no longer needed. The lander, now exposed to the Martian atmosphere, saw the red ground of Mars for the first time. A radar turned on so the robot could tell how high it was and determine when to turn on its rocket engine for the final descent.

If the lander was coming down on Earth, the parachute would gently lower it all the way to the ground. But, unfortunately, the air on Mars is much thinner, so a parachute won't slow down the capsule enough for a safe landing.

At each announcement—"Parachute deploy!" then "Heat shield eject!"—another cheer filled the room. We were all hoping for the best as the robot made its dash to the ground. With less than a kilometer to go, the lander was lowered below the parachute on a long cable. Six huge airbags were inflated, completely surrounding the lander until it looked like a giant cluster of grapes. Rocket engines attached to the cable fired to bring the lander to a stop just above the ground. Then the cable was cut, and the whole device dropped like a giant beach ball onto the ground.

"Contact!" yelled the mission controller. Everyone, scientists included, jumped up and waved their arms in the air, cheering wildly.

We'd landed on Mars!

Then the controller's voice came over the speaker again. "Okay, everyone, calm down. We've lost signal."

Oh no!

Had the lander crashed at the last second? Was the whole mission lost because of something that happened right as the rover touched the ground?

A hush came over the room as everyone thought about how sad it would be if everything was lost after so much work by so many people.

"It is bouncing on the surface. We have to wait until it stops."

Of course! That's what the airbags were for. They cushioned the impact with the ground, but they also made the lander bounce. In the low Martian gravity—which is only one-third as strong as it is on Earth—the first bounce

sent the lander flying back into the air up to the height of a four-story building. There was another bounce after that, and another and another.

Finally, after several minutes, we got the signal that the rover was safe on Mars. Cheers went up all around once again, this time for real. All eyes were glued to the television monitors as the first picture arrived—an image of the rover itself, its deflated airbags along the edge of the frame. The image proved that the rover was right-sided up and healthy. It could have tumbled over on its side or slid between some rocks, but all looked clear.

Then, as the rover lifted its mechanical head and looked out to the horizon, we saw flat ground with hills off in the distance. We were vicariously standing on another world, taking a look around. It was as if we were there with the rover, standing on Mars.

In the future, other robots will venture out and explore strange new worlds to prepare the way for human beings to follow later, just as *Spirit* did. And even when we do land on other worlds, we will likely have robotic helpers with us to assist in the work. In some cases, robots will be able to go places we can't, like into the poisonous clouds of Venus or the super-cold liquid methane lakes on Titan, a moon of Saturn.

Human astronauts are often called heroes because of the dangerous work they do. Robots do even more, but they can't appreciate where they are and what they've done.

Or can they?

YOU TRY IT!

Turn Your Friend into a Robot

Who doesn't want to turn their friends into robots? Or maybe one of your siblings or coworkers? Here's how!

WHAT YOU NEED

1. Two cell phones with video capability

WHAT TO DO

1. Move to two different locations where you can't see each other, either in another room, another building, or outside. (Tip: It's best if the person who's the robot does not tell the controller where they're starting.)
2. Connect to your friend by video on your phones. You're going to take turns, with one of you as the controller, the other as the robot.
3. The robot person holds the phone so that it points straight ahead, away from them. This way, the control person can see where the robot person is.
4. The control person gives a command to the robot to do something, such as turn to the right, move forward or

backward, or zoom in on an object—similar to commands that would be sent to a robot on Mars.

5. When you're the robot person receiving a command, count to ten before doing what your friend says. This represents the time delay for a signal to travel between Earth and Mars. (The real time delay is twenty minutes each way, but that would make for a long and boring experiment.)
6. The control person needs to tell the robot when to stop and start, but maintain that ten count between every command and execution, even if the robot is about to crash into a wall. Just know that if that happens, the mission is over.

PART 4

Weird, Wacky, and Wonderful

Strange Galactic Phenomena

24

How Did an Asteroid Wipe Out the Dinosaurs?

Sixty-six million years ago, a visitor from space landed in Mexico. It was a space rock, a flying mountain twelve kilometers across that slammed into the Earth at more than fifty thousand kilometers per hour. The explosion shook the planet, gouged a huge crater fifty kilometers wide into the coast of what is now Chicxulub, Mexico, sent enormous tsunamis racing across the Atlantic Ocean, and set fire to the Earth and the atmosphere.

As a result of the fires and the global cloud cover, acid rain fell from the sky for months after the impact. And because the asteroid kicked up a lot of dust and debris, our planet was very dark for about a year. This darkness and acid destroyed plants, which were food for the big, herbivorous dinosaurs, such as duckbills and ceratopsia, that were roving the planet at that time.

Most of the large dinosaurs—such as the T. rex, triceratops, and duck-billed dinosaurs—couldn't adapt to the new conditions. They didn't survive for long after the asteroid impact. But other dinosaurs did. Birds are direct descendants of dinosaurs, and they made it through the experience, along with a number of animals that are familiar to us today—crocodiles, turtles, salamanders, frogs, and, of course, mammals. When the dinosaurs went extinct (at least the big ones on land), mammals took over and diversified within a few million years.

It's hard to believe that creatures that had been on this planet for 150 million years (much longer than humans have been around) were wiped out by a single rock. But that wasn't the first time that life on Earth was killed by a large object falling from space. Scientists believe it happened at least four other times before that. In fact, 250 million years ago, long before even the dinosaurs existed, an asteroid wiped out almost all life on our planet, including life in the oceans. That event is known as the "Great Dying."

But life on Earth did grow back. It was just different from the life that existed before. That new wave gave way to the dinosaurs. After the dinosaurs disappeared, mammals took over, and . . . well, here we are. In other words: we owe our existence to an asteroid.

Evolution is an essential part of life on Earth. A big change in the environment—whether it's from something hitting the planet or volcanic activity—occurs every hundred million years or so. So here we are: big, dominant creatures thriving on the planet today. Does that mean we're the next to go? Is there a giant space rock just waiting to take us out?

Before you panic and go outside screaming about the sky falling, it might help you to know that we have something the dinosaurs didn't. We have telescopes. We can see asteroids coming toward us. Fortunately, none have been spotted yet that are dangerous. But if one is found, there are several ways we could steer it away so that it doesn't hit us.

We could send up a nuclear weapon and explode the incoming asteroid, provided we see it years in advance. The problem with this approach is space junk. Even if we blew up the asteroid, lots of smaller bits would still hit the Earth and do serious damage.

A better option is to gently push the asteroid just a little off its course so

it will miss the Earth. One way of doing that is to send a spacecraft out to meet it. The spacecraft would use its rocket engine to push on the asteroid and alter its course.

Another idea is to simply hit the asteroid with something heavy. We would have to be careful not to hit it too hard, of course; otherwise we might break it into pieces. We've sent out spacecraft to see what asteroids look like close-up, and we found that many of them are made of very loose material, almost like gravel, so they break apart very easily. We could hit the asteroid gently again and again until we nudge it off course. A mission called DART, which stands for Double Asteroid Redirection Test, will try out this idea on a small asteroid.

We could also use explosives to change the asteroid's course. Again, we wouldn't want to blow up one big problem into a bunch of smaller ones. But if a large bomb were detonated beside the asteroid instead of right on it, the force of the blast could nudge it off course.

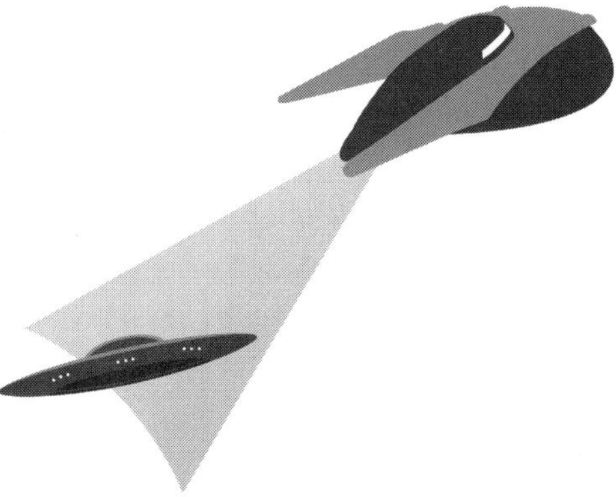

In science fiction, such as *Star Trek*, a tractor beam is sometimes used to tow a smaller spacecraft into a bigger one. No such beams actually exist, but believe it or not, the same thing can be accomplished using gravity alone. All objects possess gravity, so if a massive spacecraft was parked beside an asteroid, there would be a small gravitational attraction between the two of them. This would normally cause the spacecraft to crash into the asteroid. But if the spacecraft ran its engines and pulled *against* the asteroid's gravity and toward itself, the asteroid would be pulled toward the ship. This gravitational tug, while not very strong, would be enough to steer the asteroid off course.

Finally, one of the simplest ways to move an asteroid is to send up a big can of paint and color the whole thing white. No, really, it's not a joke! Sunlight reflecting off the white paint would push on the asteroid gently, turning the asteroid into a solar sail and altering its course.

SPACE PLACES

In a desert of Arizona, in the southwestern United States, there is a very large hole in the ground called Meteor Crater. The big circular hole, shaped like a deep salad bowl, is more than a kilometer across and was gouged out by a meteorite the size of a house that crashed to Earth fifty thousand years ago. Since the area is a desert, there has been very little rainfall to wash the crater away, so it is almost perfectly preserved. This is one of the few places on Earth where you can see how much damage an object falling from space can do to the ground.

Now imagine the destruction sixty-six million years ago, when the object that hit the Earth was the size of a mountain. You can see what the dinosaurs were up against!

The most important part of moving an asteroid off its collision course with the Earth is to reach it when it's really far away. From a great distance, the Earth is a much smaller target, so it doesn't take much of a push to make the asteroid miss us, the same way that even a small change from a pitcher's arm can mean the difference between a strike and a ball in baseball. But as the asteroid gets closer, it has to move a much greater distance to get out of the way. That takes a lot more energy. If it's too close, though, there's nothing we can do. Earth will have another bad day.

Asteroid impacts are a natural hazard that we could prevent with enough advance notice. That means that, for what may be the first time in our planet's history, we can prevent an asteroid apocalypse. Too bad the dinosaurs didn't have that option.

YOU TRY IT!

Collision Course

The Earth has been hit by large objects from space more than once, which means it could happen again in the future. Fortunately, we have telescopes looking out for any that may be heading our way. Here's how you can see why it is important to get to them early and nudge them off course.

WHAT YOU NEED

1. A globe of the Earth, or a large ball
2. A handful of small pebbles or marbles

WHAT TO DO

1. Stand beside the globe or ball and try to hit it by throwing a pebble.
2. Take five steps away from the ball and try to hit it with another pebble.
3. Take another five steps back and try again. Continue stepping back until it is almost impossible to hit the ball.

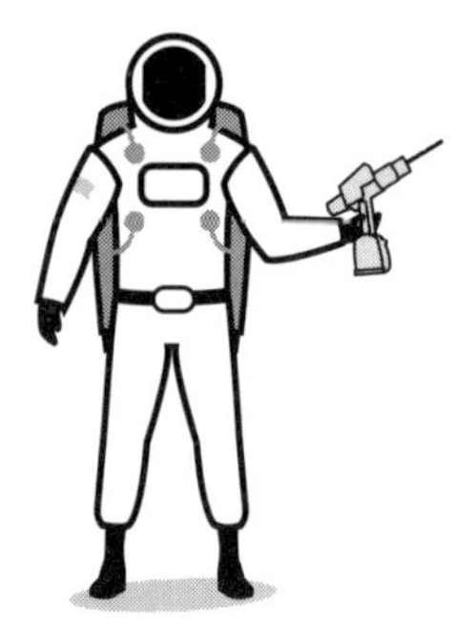

It's easy to hit the ball when you are close because it's a big target. But as you get farther away, the ball looks smaller and is more difficult to hit. From really far away, your aim only needs to be off by a tiny bit and you will miss.

If we can send spacecraft out to meet an asteroid when it is far away from the Earth, it would only take a small push to change its course enough to ensure it passes by our planet without hitting us. If we wait too long, or don't see an asteroid until it is closer, though, it would have to be moved much more to steer it out of the way; otherwise the Earth would be an easy target. And if it's too close, there will probably be nothing we can do except get ready for impact and hope we don't go the way of the dinosaurs.

25

Do Other Planets Have Weather?

Almost all the planets in our solar system have weather. The only one that doesn't is Mercury, because, like the moon, it doesn't have any air or atmosphere. After all, that's what weather is: moving air. But not all planets have the same type of air that we have here on Earth, so weather on other worlds can be quite different.

We live on a very active planet. Our air is always moving. It swirls around the globe, stirring up the oceans. It blows water, snow, and sand around with such incredible force that it can wipe out just about anything we can build. Tornadoes rip houses apart; hurricanes do incredible damage to entire cities. All that damage is done by air, the stuff you're breathing right now.

Venus, the second planet from the sun, has much more air than the Earth, with clouds so thick we can't even see through them. Mars, planet number four, has much less air, but it has very strong winds that produce enormous dust storms that occasionally cover the whole planet. Then there are the gas giant planets—Jupiter, Saturn, Uranus, and Neptune—which are nothing but weather. These huge gas worlds, much farther from the sun than Earth, swirl constantly in endless storms. There is no land, only clouds, and they're always moving—the weather is always bad. Clouds on Jupiter are thousands of kilometers tall and come in different colors—white, blue, and red. That's because they're made of different chemicals, including ammonia and hydrogen sulfide, which smells like rotten eggs.

To give you an idea of just how bad the weather can be on another planet, the Great Red Spot on Jupiter is a storm that has been raging nonstop for as long as we've had telescopes to observe it, which is almost four hundred years. This storm was spotted by Galileo when he pointed the first telescope on Jupiter in 1610. The storm is so big it could swallow the entire Earth with room to spare. It makes our hurricanes look tame.

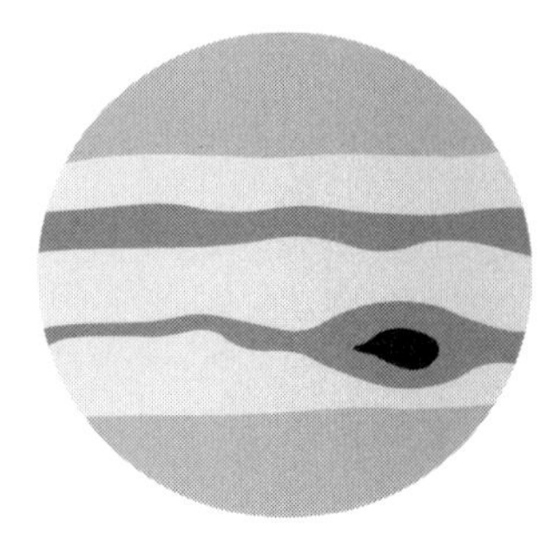

The Earth is heated by the sun, and the sun shines most powerfully on the middle of our planet. As the Earth goes around the sun every year, the North Pole always faces up and the South Pole always points down. That means the sun shines mostly on the equator, and so the middle of the Earth gets more heat than the top and bottom of the planet. This warm air around the equator tries to move toward the cold regions, and as it pushes outward, the cold air at the top and bottom tries to come toward the equator to replace it. And if that isn't enough, this air is moving around a globe that spins on its axis every day, which really mixes things up!

The same principle of heat works on other planets, except those worlds are either closer to or farther away from the sun than we are, and the gases that make up the air are different. Take Venus, for example. The weather on Venus is pretty gloomy. It's very hot and totally covered in clouds. There's not much sunlight down on the ground because so little gets down through the clouds. The clouds themselves are made of sulfuric acid, which would burn your skin when it rains. If it rains at all.

Venus, where it rains sulfuric acid.

You can't breathe the air on Venus because it's made of carbon dioxide, which is a greenhouse gas that traps heat. The average temperature on Venus is more than 450 degrees Celsius. That is as hot as a pizza oven. And it doesn't matter if it is day or night, summer or winter—the temperature never changes. It's a nasty, nasty place. You don't want to go there for a summer vacation.

What happens when we go farther away from the sun, though? The next planet out from Earth is Mars, which doesn't have very much air, but it does have weather. Mars's atmosphere is made of carbon dioxide just like Venus's, but on Mars, that atmosphere is very, very thin. It would be like being on the top of two Mount Everests—very little air to breathe and exceedingly cold.

Even though it's so cold on Mars, the planet still has four seasons. A summer day on Mars may get to zero degrees Celsius at the equator. The planet has ice caps at the North and South Poles and winds blowing on the surface. White clouds fill valleys in the early morning, and in the afternoon, when the sun heats the surface, the winds start playing with the dust on the ground. One of the robot rovers that landed on Mars in early 2004 captured images of dust devils—little tornadoes dancing across the desert. All that dust blowing around makes the sky on Mars orange instead of blue.

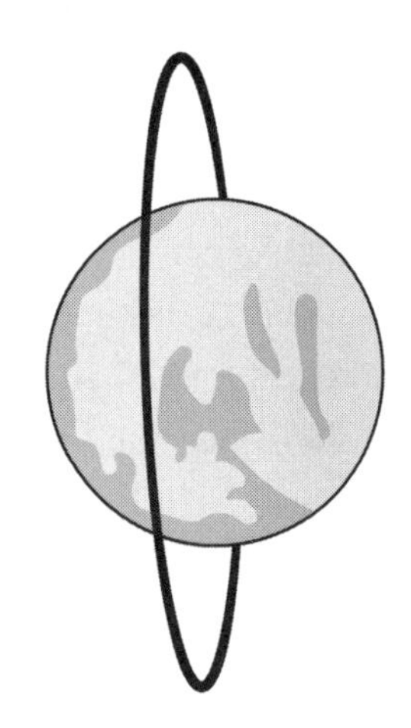

So these three planets—Venus, Earth, and Mars—are all quite different. One is too hot to have much weather, the other is too cold to have extreme storms, and one is just right. Aren't you glad we live here?

Things get even weirder the farther out in our solar system you go. Uranus is tilted on its side (for what reason, nobody knows)—so for a quarter of the year, its North Pole is aimed at the sun, then for another quarter, its South Pole is aimed at it. And between those times, the sun shines on the equator.

Sound a bit confusing? Hold a pencil in front of your face with the point up. The point represents the North Pole of a planet; the eraser at the bottom is the South Pole. Most of the planets in our solar system have their North Poles pointed in the same general direction we call north. The Earth's pole is tilted a little, but still more or less upright.

Now turn the pencil sideways to the right, so the point faces a wall. That is the way Uranus is positioned, lying on its side. If your head is the sun, the North and South Poles of Uranus always face the same way in space as it goes around in its orbit. Move the pencil to the left side of your head, and the pencil's point, or North Pole, will be aimed right at you. Move it around to the right side of your head, and the end with the eraser, the South Pole, will be pointed at you.

This makes for extreme seasons on Uranus, where the top of the planet is heated by the sun for half of its year and the bottom for the other half. And a year on Uranus is eighty-four Earth years long!

Some of the weirdest weather isn't found on planets, though—it's on their moons. Titan, a moon orbiting Saturn, is one of the weirdest worlds of weather anyone has ever seen. It's one of the largest moons in the solar system, and it has a cloudy atmosphere. A robotic probe called *Huygens* landed on Titan in 2005. It passed through brown clouds, was blown by winds, and touched down on a surface that looked wet. It rains on Titan, but it's nothing like the rain we get on Earth.

Titan is extremely cold, hundreds of degrees below zero. When it's that cold, water is frozen solid. In fact, the ground on Titan is made of ice that's as hard as rock. Titan's surface is covered with river valleys and what look like shorelines and beaches. There are very large lakes, too—one of them is as large as Lake Superior in Canada. The lakes are filled with rain, but it's not made of water. Because on Titan, it rains liquid methane, the same stuff we use in gas appliances in some homes.

Pluto is also an ice world with a very thin methane atmosphere. In fact, Pluto is so far from the sun that occasionally its thin methane atmosphere completely freezes, forming snow that falls to the ground. But the snow isn't frozen water. It's frozen air.

Imagine the wonders you would see, if only you could brave the bad weather on other planets. You'll need some special gear and a spacecraft, but what a ride!

YOU TRY IT!

Weather in a Glass

Find out how different layers form in the atmosphere and how they interact with one another to make weather.

WHAT YOU NEED

1. A large glass jar
2. A small glass that fits inside the bigger one
3. Food coloring
4. Tongs
5. Warm and cold water

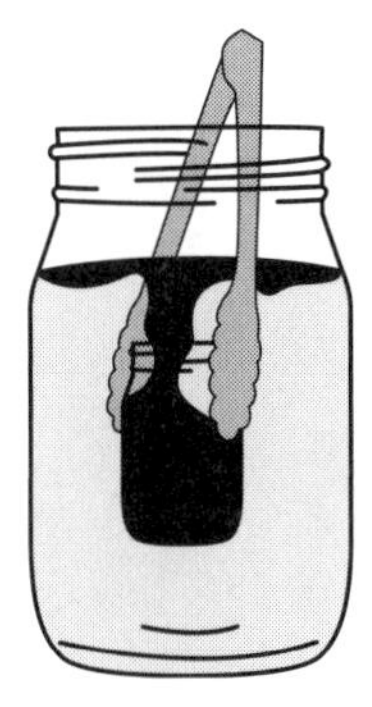

WHAT TO DO

1. Fill the large glass jar three-quarters full of cold water.
2. Pour hot water into the second, smaller glass, filling it right to the top.
3. Add some food coloring to the hot water.
4. Using the tongs, carefully lift the small glass without spilling any of the water and gently lower it into the larger glass. Make sure the small glass of hot water completely sinks under the cold water.
5. Watch what happens to the hot water.
6. You have just made a storm in a glass.

You should see that the warm water rises out of the small glass and floats to the top of the larger one. It doesn't mix with the cold water at all. That's because warm water is lighter than cold, so it floats. Air does the same thing. When warm air runs into cold air, it is pushed up by the cold air flowing under it, just like you see in the glass jar. When the warm air rises, it cools, and moisture in the air condenses into drops. The drops grow until they are too heavy to stay in the air, and the result is rain. This is the basic principle behind weather and storms, both here on Earth and across the galaxy.

26

What Are Gravitational Waves?

There are water waves that make water move, sound waves that move the air, earthquake waves that shake the Earth, and, well, gravitational waves make space move.

We know that gravity works by curving space. Large objects bend space inward to form a well, like a bowling ball pressing down on a mattress, and they draw other, smaller objects close by either falling down into the well or orbiting around the larger object. That's pretty amazing by itself, but consider this: if space can bend like that, then it means that space can also vibrate like a drum.

How do you vibrate space, you ask? With black holes!

Black holes have a lot of gravity, which means their gravity wells—the area close to the black hole where gravity gets stronger and stronger the closer you get—are really, really deep. Imagine two black holes that approach each other. As they get close, they will fall into each other's gravity well, forming a single, deeper one. The black holes will spin around each other like two figure skaters, twirling faster and faster until they finally join

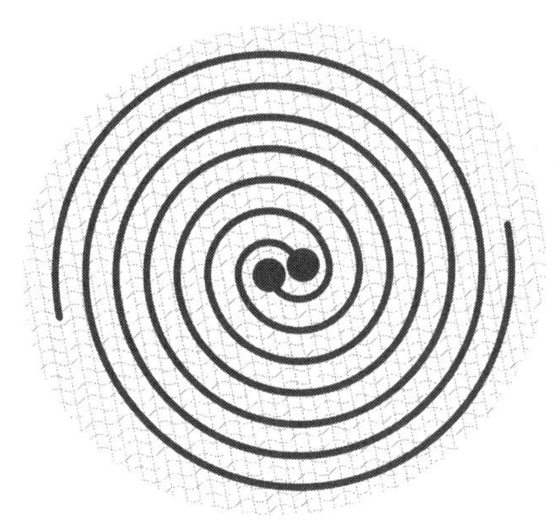

together. That spinning actually vibrates the space around them, sending waves of space itself out across the universe, like ripples in a pond.

Think of it like the surface of a drum. When you beat a drum with a stick, the surface of the drum bounces up and down, vibrating the air above it. As the air vibrates, sound waves travel away from the drum. Those sound waves are technically called compression waves, because the drum squeezes and releases the air, making little pulses of pressure. When those pulses hit your eardrums, they are pushed in and out at the same beat, and we hear the waves as sound.

Gravitational waves are also compression waves, but instead of compressing air, they squeeze space itself. Gravitational waves actually pass through us all the time. Fortunately, they're very small, so we don't feel them. But if a bigger one passed through you from head to toe, you would get a little taller, then a little shorter as the wave went by. And if it came from the side, you might get a little fatter and then skinnier. It's similar to the way that water waves make you bounce up and down while swimming.

It turns out that gravitational waves vibrate at about the same frequency as sound waves. Lucky for us, because if we can hear gravitational waves, it means we can hear the sound of space itself!

To listen to the sounds of space, we use an extremely sensitive instrument. And it doesn't look anything like an ear.

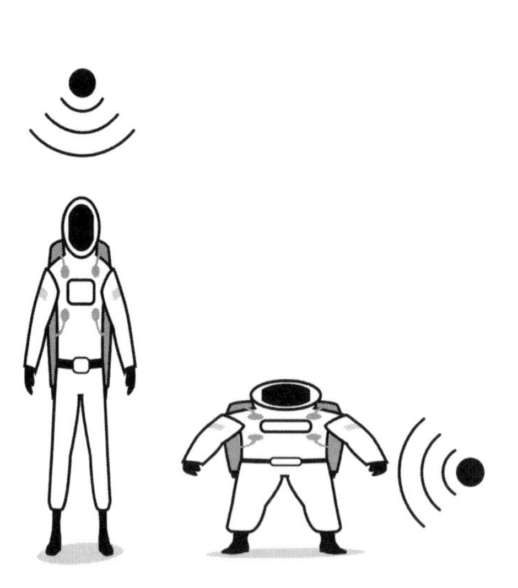

The Laser Interferometer Gravitational-Wave Observatory (LIGO) is made of two arms, each four kilometers long, that are arranged in an L shape. Inside each arm is a laser beam that shoots down the length of the hallway and reflects off a mirror at the end, allowing the technicians to measure the exact length of each arm.

When the beams bounce back to the base of the arms,

ON THE DRAWING BOARD

Some scientists believe it may be possible to design a new type of spaceship that doesn't actually fly *through* space, but instead warps space around the ship using gravitational waves. It's the theory behind the fictional warp drive that enables the crew of the starship *Enterprise* in *Star Trek* to hop around the galaxy with ease. The idea is that a ship could squeeze the space in front it so that the distance to its destination is shorter. At the same time, it would make the space behind the ship larger, so that it was farther away from where it started. That way, the ship isn't actually moving—it's the space around the ship that's changing.

A ship powered by a warp drive wouldn't even need seat belts, because the people inside wouldn't feel as though they were moving. A warp drive is, as the name suggests, warping gravity, and gravity acts on all objects at the same time. So, when space is distorted around a starship, the ship, the people, and all the objects in it—every atom—are pulled at once, making sure there's no kick-in-the-pants jolt from acceleration. It's a long way from today's astronauts, who are pushed back into their seats with a force that makes them feel three times their own weight as the rocket blasts into space.

Unfortunately, no one knows how to warp space enough to move a ship. The gravitational waves that we've detected so far are extremely small, so it would take a huge number of them to go anywhere. Maybe someday we'll figure it out, and then imagine where we might go!

they are combined together in a light detector. Lasers are made of light waves, and so long as the lasers don't move, all of the light waves line up with one another when they pass through the instrument.

Remember, though, that when a gravitational wave passes through an object, it can bend or distort its shape. When gravitational waves pass through the Earth, then, they cause one arm of the observatory to get a little longer (and then a little smaller) than the other. That makes the mirrors at the end of each arm vibrate. The gravitational waves are extremely small,

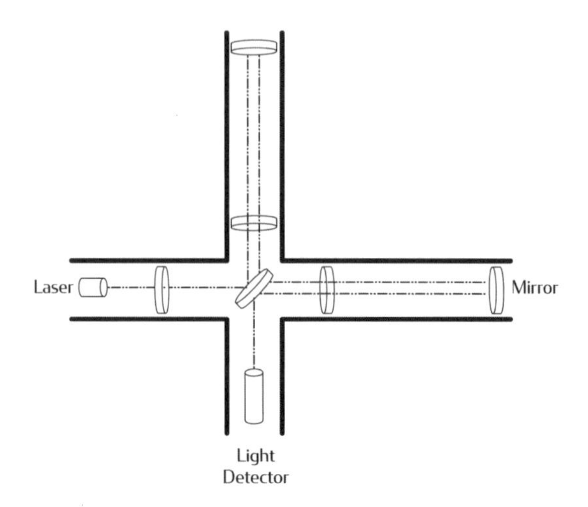

so the amount the mirrors move is less than the width of an atom. The light detector measures the light waves of each laser as they shift in and out of alignment, making a pattern in the light that vibrates and allowing us to "see" the gravitational waves.

SPACE PLACES

There are two LIGO detectors in the United States: one in Washington State, in the western part of the country, and the other in Louisiana, in the south.

In September 2015, the LIGO observatories both made the first detections ever of gravitational waves when they recorded two black holes colliding. Not only did they see the gravitational waves on their instruments, but they also turned it into sound waves so we could hear the event, too. It turns out that when two black holes collide with each other, they go "WHOOOOOP"!

Since then, scientists have seen other gravitational waves from colliding neutron stars, which are super-dense stars with powerful gravitational fields.

Scientists are excited about this new technology because it opens up a whole new field of gravitational wave astronomy. These waves travel across the entire universe and can tell us about some of the more extreme events in the universe, such as the collision of two black holes, super-dense neutron stars, or maybe even the nature of gravity itself. And since gravitational waves pass through everything, they are not blocked by dark clouds of gas in space the way light is, so we will be able to see farther out into space, and farther back in time, than ever before.

Gravitational waves are like a new set of eyes on the universe. Who knows what we will see?

YOU TRY IT!

From Sound to Sight

WHAT YOU NEED

1. A funnel
2. A piece of aluminum foil
3. A flashlight

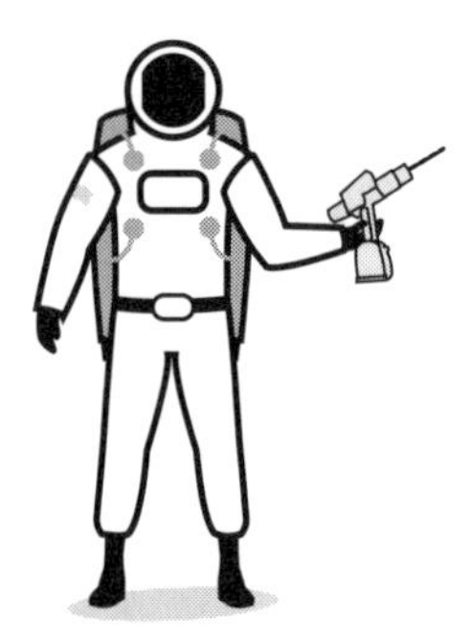

WHAT TO DO

1. Take a piece of aluminum foil that is larger than the wide part of the funnel and wrap it tightly over the top so it looks like a drum. Make sure the shiny side is facing outward.
2. Stand about a meter away from a wall and hold the funnel like a trumpet, with the aluminum side facing toward the wall.
3. Shine the flashlight onto the aluminum and turn the funnel slightly until you see a circle of light reflected on the wall.
4. Carefully place your mouth over the small end of the funnel and hum. Watch what happens to the reflection on the wall.
5. Experiment with different sounds, high and low. Try humming a tune and see how the reflection changes.
6. Congratulations, you've made your own mini-LIGO!

When you hum, you make the air inside the funnel vibrate. That makes the surface of the aluminum foil vibrate, which leads the light waves bouncing off the aluminum to also move. Low sounds are longer waves, so the light should vibrate more slowly, compared to high sounds that vibrate more quickly. If you couldn't hear the sound, you would still know whether it is high or low by looking at the speed of the vibrating light. Astronomers can tell the same thing about the nature of gravitational waves by the vibrations of the light waves in LIGO.

27

Why Isn't Pluto a Planet?

Pluto used to be called a planet. Now it's called a dwarf planet. So what happened?

When you hear the word "planet," you probably think of Mars, Venus, or the Earth—planets that are made of rock. Maybe you think of giant Jupiter or beautiful Saturn—huge balls of gas hundreds of times bigger than Earth. You might even imagine the bluish-green worlds of Uranus and Neptune, which are also made of gas.

But the oddball planet out there beyond Neptune is a tiny, frozen world—Pluto! Named after the Roman god of the underworld, it's a dark place, so far away that the sun is just a small, distant star in the sky. Daytime on Pluto would seem like early evening just after sunset on Earth. That also makes Pluto cold—colder than anyplace you can imagine.

Pluto, a dwarf planet

This little ice world is very small. You could drop the whole planet onto the Canadian prairies and it would fit between the cities of Calgary and Thunder Bay. And it's very far away. All the other planets, including the Earth, go around the sun in almost perfect circles. But Pluto follows a

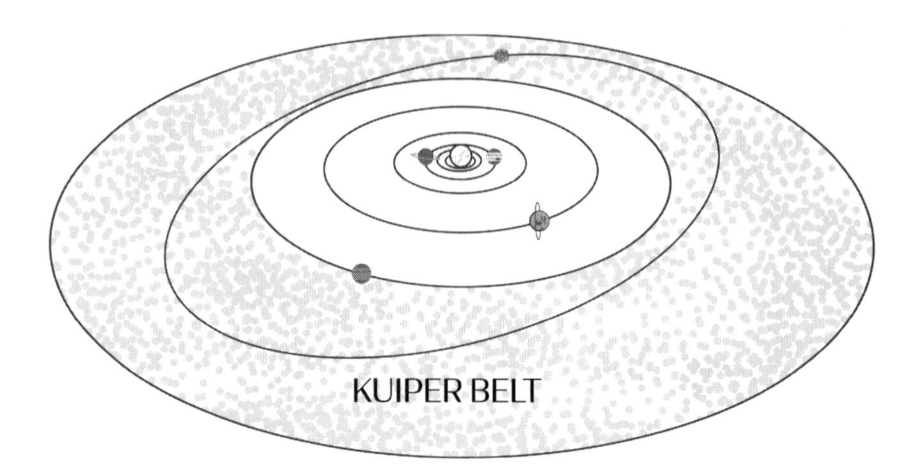

noticeable ellipse, which is sort of a football-shaped orbit. It's also a very big orbit. One year (or one orbit) on Pluto is 248 years on Earth. In fact, Pluto hasn't even gone around the sun once since we discovered it.

Not only that, Pluto's football-shaped orbit is so lopsided it crosses the orbit of Neptune. Don't worry, they won't run into each other because the other weird thing about Pluto's orbit is that it's tilted on an angle—it goes above and below the other planets, as well as in and out.

When Pluto was discovered about ninety years ago, it was thought to be the only planet of its kind. But since then, many other icy planets very much like Pluto have been found in the same region of space. That's why people are saying Pluto shouldn't be called a planet at all.

There are thousands of snowball objects surrounding our solar system in a huge swarm called the Kuiper Belt, named after Dutch American astronomer Gerard Kuiper. These ice worlds are believed to be leftover stuff from the giant cloud of gas and dust that formed all the planets and moons billions of years ago. That makes them really interesting to study because they haven't changed much in all that time. Looking at these snowballs is like looking at the past, to a time before the sun was even born.

Pluto sits right in the middle of the Kuiper Belt. If Pluto remained a planet, we would have to classify all the other icy objects in the Kuiper Belt as planets as well. Can you imagine trying to remember the names of thousands of planets? That's why Pluto was renamed a dwarf planet, which means a celestial body that still goes around the sun and may have moons, but is small and part of a swarm of other, similar objects. We can think of Pluto as king of the dwarves!

Most planets are fairly easy to spot in the night sky. They tend to be a little brighter than the stars, and if you watch them for a period of time—say, over a couple of weeks or a month—they change their position. They seem to wander among the stars, which is what the word "planet" actually means "wanderer."

But Pluto, small and so much farther away than the other planets, is not very bright. It looks just like a dim star, even in a telescope. And because it's so far away from the sun, it moves very slowly, which is why Pluto was the last planet to be discovered. And doing so wasn't easy.

Can you tell the difference between those two photos below? Notice the little arrows—they point to the only dot that is in a different position.

That is exactly how a young astronomer named Clyde Tombaugh discovered Pluto in 1930. He used a small telescope at Lowell Observatory in Arizona to take pictures of the sky, then he compared his photos to old pictures of the same areas of sky that had been taken before. While flipping back and forth between old and new pictures, he saw what looked like a star that had moved. It took Clyde about seven thousand hours before he finally spotted the tiny little wanderer beyond Neptune, now known as Pluto.

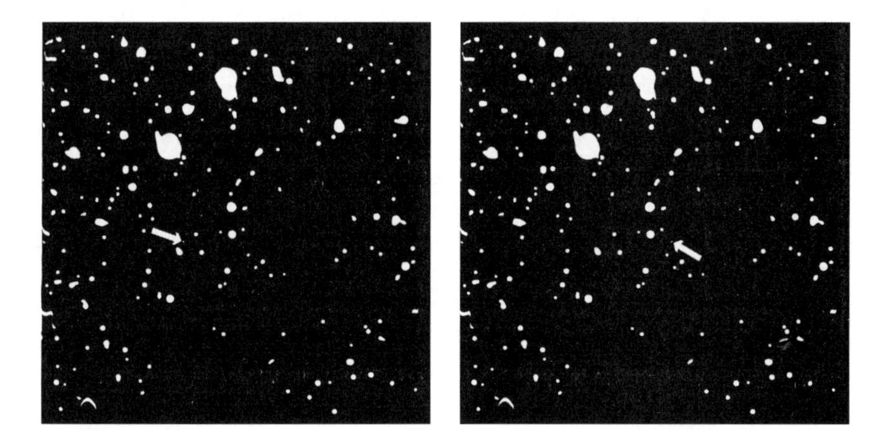

Only one spacecraft has ever visited Pluto. The *New Horizons* probe, about the size of a piano, took nine years to cross the solar system, then whizzed by the dwarf planet in one day in 2015. But what a day it was!

It's cold out there on the edge of the solar system, six billion kilometers away from the sun. A warm day on Pluto is about 230 degrees Celsius—below zero! The *New Horizons* probe confirmed Pluto's small size. The planet is only about two-thirds the size of our moon.

Pluto has a moon called Charon that's half as big as it is. Usually, a moon goes around a planet, but Charon is big enough that both it and Pluto go around each other like a pair of figure skaters. So, in a sense, Pluto is actually a double planet, with two parts to it spinning around each other. Both Pluto and Charon are made mostly of ice—two snowballs circling each other in deep space. Four other little moons around Pluto—Hydra, Styx, Nix, and Kerberos—are much smaller and look like they were chipped off of something bigger. Maybe they were!

Pluto has an atmosphere, but you would choke if you tried to breathe the air. It is made of methane, which is a natural gas we also have on Earth. Imagine air made of the same stuff that cooks hot dogs on a gas stove. The atmospheric methane gives the air on Pluto a blue color. Under that blue sky are steep-sided mountains as high as the Rockies, but they're made of ice—three kinds, to be exact. There is frozen water, like we have here on Earth. That forms the ground and mountain peaks. Then there's methane ice that freezes out of the air. When it first forms as frost, it's bluish white, but after the sun shines on it for a while, it turns reddish brown. Finally, there's nitrogen ice.

Never heard of it?

Take a deep breath. You just breathed in nitrogen. That's right, our air is mainly nitrogen gas with some oxygen in it. On Pluto, it's so cold, our air would freeze and turn into snow. But don't try to make a snowman out of nitrogen snow. It won't stay together. Even though it's white and frozen, it moves like thick pudding. A snowman would quickly drip down and become flat. That's why there are huge flat plains of nitrogen ice all over Pluto's South Pole.

After passing Pluto, *New Horizons* continued farther out into the Kuiper

Belt and passed by a much smaller icy world called Ultima Thule, which turned out to be shaped like a snowman! It's a strange place out there at the edge of the solar system. The spacecraft will eventually leave our solar system altogether to wander among the stars of the Milky Way for billions of years. Its trip has been worth it, though—after Pluto turned out to be such an interesting world, some astronomers suggested it be turned back into a full planet.

The problem of whether Pluto is a true planet or not comes down to how we define what a planet is. We used to think planets were like the Earth—big balls that orbit the sun. Then we found that there are also small things going around the sun, and we called them asteroids and comets. Then we found Pluto and discovered it's a snowball and there are lots of little snowballs out there with it. So what do we call it? Is it a planet or something else?

Think of it this way. Collect a bunch of different rocks and lay them out according to size. Now, which ones do you call rocks, which ones do you call pebbles, which ones do you call stones, and which ones are boulders? When objects are different sizes, each of us has our own idea of what to call them. And that's the issue facing astronomers.

Astronomers think there could be two hundred dwarf planets wandering around our solar system. Pluto is one of four named dwarf planets

among the Kuiper Belt of icy objects. But others have already been spotted out in the same region and given names, such as Sedna, Quaoar, and Eris (which is actually larger than Pluto). New dwarf planets are being discovered all the time, so there are hundreds of others out there that have yet to be named.

And not all of them are among the snowballs beyond Neptune, either. One dwarf planet, Ceres, rests in the asteroid belt that sits between Mars and Jupiter. Found in 1801, Ceres is simply a really big asteroid, big enough to be considered part of the family of dwarf planets, which makes those who study it happy.

So what's the final word on when to call something a dwarf planet versus a fully fledged planet? Anything in the solar system that is very large, has pulled itself into a round ball, and that follows its own orbit around the sun should be called a planet. But if it's small and round and part of a group of other objects that share a path around the sun, it's a dwarf planet.

Clearly, our definition of planet has changed a little bit over the years to acknowledge both the big ones and little ones, and dwarves are the new kids on the block. It might seem a little confusing to have names change, but that's how science works. As we learn new things, we add them to our knowledge base. In space science, we're still learning about our place in the universe. There's a whole lot out there that we haven't discovered yet, so let's get into space and find out more.

YOU TRY IT!

A Sweet and Salty Kuiper Belt

Building a model of the solar system helps us see just how big the Kuiper Belt is.

WHAT YOU NEED

1. One grapefruit
2. Two marshmallows
3. Two Smarties
4. Four poppy seeds or tiny pebbles
5. Sugar or salt

WHAT TO DO

1. This is best done on a dark table, desk, or countertop. Place the grapefruit on the table to represent the sun. Place the seeds around the grapefruit, each one a little farther away to represent the rocky planets Mercury, Venus, Earth, and Mars.
2. Place the two marshmallows farther away to represent the giant gas planets Jupiter and Saturn. Two Smarties represent the big planets Uranus and Neptune at the outer edge of our solar system. (The planets do not have to be in a straight line. That rarely happens in space.)
3. Finally, sprinkle sugar all the way around your solar system model in a ring that is larger than the distance Neptune is from the grapefruit.

Congratulations! You have just made a (delicious) model of the solar system! The sugar represents the Kuiper Belt. Now try to pick up just one grain of sugar or salt on the end of your finger. That's Pluto.

Also, keep in mind that this model is only showing the size of the planets, not the distance between them. If we were to spread them out to the full scale of our solar system, the sugar would be several blocks away from the grapefruit.

28

Why Do Stars Twinkle?

Stars don't actually twinkle. Astronauts in space look out into the night and see stars that are perfectly still—no twinkling, and no change in their light. It's only when we look up at the stars from the ground that we see them shimmering. That's not because the stars themselves are suddenly winking at us—it's the air above us that is moving.

On clear, dark nights, when the stars shine like diamonds in the blackness, it's easy to forget that we are gazing through a thick layer of air that is always over our heads. The Earth's atmosphere is like a window we must look through to see out into space.

Unfortunately, our air window is not perfectly clear. Just as the windows of a car or bus are hard to see through when it is raining or when the windows are dirty, our air carries moisture, which we sometimes see as clouds, along with dust, and, sadly, pollution.

Air is also moving all the time, which is why we feel wind, and some air is hot, so it rises. On really hot summer days, if you look down a long road or a city street a block or two away, you might see the pavement shimmering. That's the air rising up, heated by the hot pavement. As it moves, it distorts the light coming to your eyes so the road seems to be moving when it really isn't.

You can also see the shimmering air effect if you're sitting around a campfire and look at people on the other side of the flames. Their faces will seem to be moving because of the distorting hot air rising off the fire.

All of that movement and congestion also gets in the way of starlight. Our eyes are seeing starlight after it has passed through that air junk and activity to reach us. As the starlight streams down from space, the moving air interferes with the light, which causes the shimmer effect that we see as twinkling.

Twinkling stars look nice. We even write songs about them, like "Twinkle Twinkle, Little Star." But for astronomers trying to see the stars through telescopes, that twinkling is a big headache because all that shimmering gets in the way of taking clear pictures. The stars we see from Earth are moving around, so images in our cameras are always a little blurry. That's why big

telescopes are usually on top of mountains. Up that high, they're above a lot of the moisture and pollution, so the stars don't twinkle as much.

Of course, the best way to bypass the twinkling effect is to put a telescope in space, where it's completely above the air. From that vantage point, the stars will not twinkle at all. That's why the Hubble Space Telescope takes the clearest pictures of any telescope. It's not because it's the biggest telescope in the world. The mirror on the Hubble telescope is only 2.4 meters (94.5 inches) across. On Earth, some telescopes are being built with mirrors more than thirty meters (1181.1 inches) across. Hubble had to be smaller so that it could fit inside the cargo bay of the space shuttle for launch. But because it doesn't look through air, its images are stunning. Only the megatelescopes of the future will be able to beat it, and they need a system called adaptive optics to equal the quality of the relatively small Hubble.

There's one clever way astronomers take the twinkle out of the stars: they make the telescope twinkle, too! Well, not the whole telescope—just a tiny mirror that catches the starlight before it gets to the telescope camera. (Remember, telescopes are just cameras with really big lenses.)

Here's how it works. After the starlight has gone through the telescope, but before it gets to the camera, the light hits another small mirror that moves just like the stars do. It does so by shooting a laser beam into the sky in the same direction the telescope is looking. The beam shines on the upper atmosphere of the Earth, where it is moved back and forth by the air. As the air moves the beam around, the beam guides the little mirror to move in the same way and at the same speed as the air. That keeps the starlight steady in the little mirror, which then sends it on to the camera. The whole system together then takes the twinkle out of stars so their images are much clearer.

Your eyes do this all the time when you see moving objects, such as a ball that is bouncing toward you. Your eyes move up and down with the ball so that you can keep it in focus and make the catch. If your eyes didn't move, the ball would be blurry . . . and you would get hit in the head.

When it comes to telescopes, size matters. The bigger the mirror, the farther into the universe the telescope can see. While telescopes on the ground are becoming gigantic, there is also a bigger one heading to space called the James Webb Space Telescope. It has a mirror made of eighteen segments that fold up so that the whole thing will fit into the nose cone of a rocket. When it reaches space, the telescope will unfold like a flower in space to make one single mirror that's 6.5 meters across—more than twice the size of Hubble. Out in space, with no air to look through, the Webb telescope will see even farther into the depths of the universe than we ever have before.

YOU TRY IT!

The Twinkle Test

Not all stars twinkle the same. Some twinkle more than others, and some are more colorful depending on where they are in the sky.

WHAT YOU NEED

1. A clear night when you can see the stars, even in a city
2. Your own eyes
3. A pair of binoculars (optional)

WHAT TO DO

1. Find a place where you can see the stars all the way from straight overhead down to the horizon. Watch the stars straight above you and try not to blink (this works better with a pair of binoculars). Notice how much they twinkle.
2. Now find a star that is really low in the sky close to the horizon. Watch it carefully for a few moments and see if there is a difference in how much it twinkles compared to the star overhead.

The star near the horizon should twinkle more because you're looking through more air in that direction. The more air you look through, the more it obstructs and moves the starlight.

You might also find that the stars close to the horizon are more colorful. Air stops some colors of light more than others. Red light goes through air better than blue, which can give stars a pink color. That is also why the sun turns red at sunset—blue light is being stopped while the red gets to your eyes. It also makes it more beautiful, don't you think?

29

What Are Shooting Stars?

Despite what their name might say, shooting stars are not really stars. They're actually tiny bits of sand, pebbles, rocks, and boulders that fall from space and shine like stars for a second or so as they burn up in our atmosphere.

Space is not as empty as it looks. Our solar system started out as a nebula, an enormous cloud of gas and dust. Most of the cloud became the sun and the planets that we know today, but there were a lot of little bits left over that are still floating around. That's what the Earth runs into as we journey around the sun every year.

Let's not forget that our planet is speeding along through space at one hundred thousand kilometers per hour as we circle the sun every year. We don't feel that motion because we're carried along with the Earth and the ride is super smooth. The same thing happens in an elevator. Once the doors close and it gets going, it's hard to tell you're moving between floors because it runs at a constant speed. You only feel the motion when that speed changes or the elevator stops.

Every second, we travel thirty kilometers through space. Think of a place that is thirty kilometers away from you right now. (You may have to look at a map.) Then count, one-two. That's how long it would take to get there if you could move as fast as the Earth moves through space.

Our speedy planet plows through all that dust and dirt that lies in our path, like a car running into raindrops when driving through a storm. And

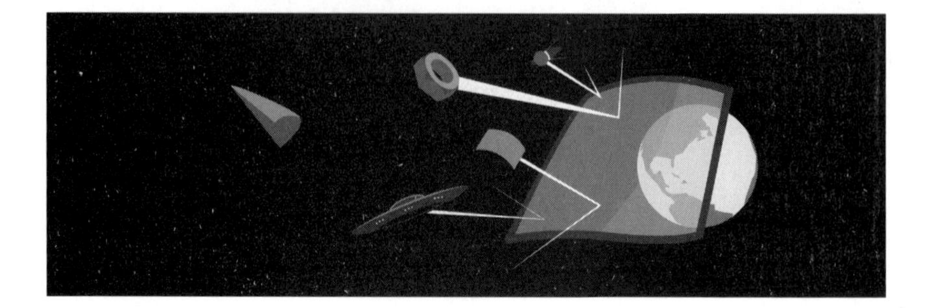

in the same way that rain hits a car's windshield, the dust and dirt floating around in space hit our planetary windshield: the Earth's atmosphere.

The particles are not just lying still out there. They're also moving extremely quickly, so they hit the atmosphere traveling many times faster than bullets. At that speed, friction from the air heats them to high temperatures, and, for a second or two, they shine as bright as stars while they burn up and shoot across the sky. That's where they get the name "shooting stars."

Shooting stars are properly called meteors, and most of them are the size of a grain of sand. They're so small that they burn up completely in the air and never reach the ground. There are plenty of them, too—the Earth picks up about one hundred tons of space dust every day.

Some meteors are a little bigger, the size of your fist or a loaf of bread. Let's call them space rocks. Space rocks can make it all the way to the ground, and when they do, we have a new name for them: meteorites. A good way to remember the difference between meteors and meteorites is that meteors don't reach the ground, but meteor*ites*, *might* hit you on the head . . . though if one hits you on the head, it would be a meteor*wrong*!

Then there's the really big debris. Some of the objects floating in space are the size of houses, others as big as a mountain. These are asteroids and comets. It doesn't happen very often, but if one of those big ones were to hit, it would be a problem, as the dinosaurs found out sixty-six million years ago.

Thankfully, we can still enjoy close-up views of some of these space projectiles without the risk of going the way of the dinosaurs. The best time to see meteors is during meteor showers. One of the best meteor showers

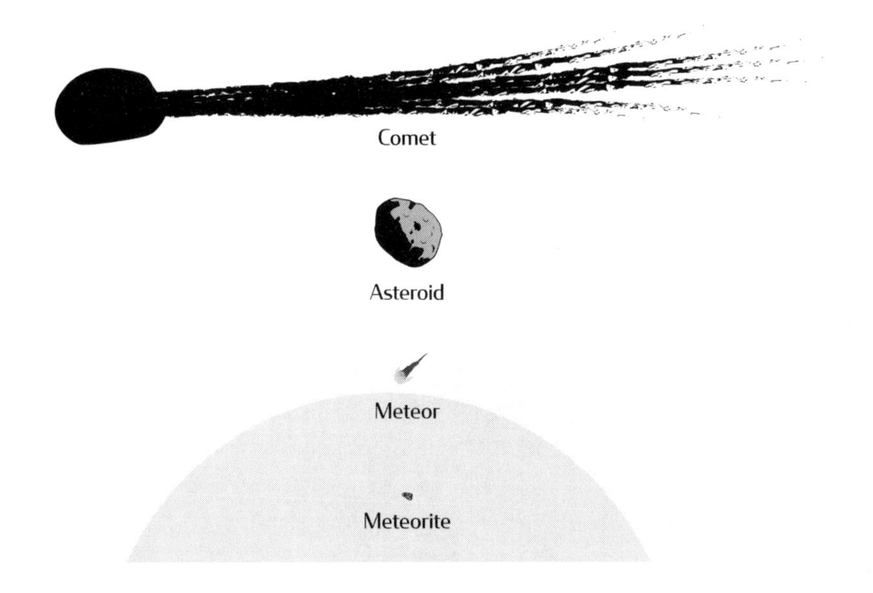

happens every year during the second week of August. It's called the Perseid meteor shower, and it occurs every summer when the Earth plows through the same path of dirt and ice in space that was left behind by a comet.

Comets are big boulders the size of mountains. They're different from asteroids because they also contain a large amount of ice. When comets get close to the sun, the ice warms up and turns into a gas that, along with a lot of dust, blows off the comet, forming a beautiful tail. Comet tails can be millions of kilometers long. The dust they give off stays in space, just like the dust cloud left behind by a big truck driving along a dirt road.

The dust that forms the Perseid meteor shower trails from a comet called Swift-Tuttle, which crosses the Earth's orbit around the sun. Every August, we pass through that leftover comet tail and get a light show of meteors.

Catching a glimpse of this celestial show is an activity best done with a group of friends. Check newspapers or the NASA website to find out what time in August is best for seeing the shower. (It is usually around the twelfth or fifteenth.)

Pick a clear night in that time when you can see the stars, then head to a place as far away from city light as possible. You will have to stay up till around ten p.m., when it is dark. Lay back on the ground or sit in a lawn chair facing east and look up. Try not to blink too much because meteors are fast and easy to miss. Be patient. Once every minute or so, you will see a quick flash in the sky, a streak of light that lasts less than a second. Once in a while, you might see one that has a glowing head and long tail behind it. Those are the bigger meteors, the ones the size of pebbles. You might even catch the odd satellite passing overhead as a bonus.

When you see how fast those meteors are moving, remember, it's not just their motion that causes the brilliant streaks across the sky. It is also you, zooming through the cosmos on your very fast spaceship: Earth.

YOU TRY IT!

Moon Meteors

The Earth is not the only planet that gets hit by meteors. All planets do, as does the moon. Look at the face of the moon and you will see it is covered in round craters of all sizes. Those craters were made by meteors large and small that have been hitting the moon for billions of years.

You can make your own moonscape of craters.

WHAT YOU NEED

1. A large bowl
2. Flour
3. A glass of water

WHAT TO DO

1. Pour the flour into the bowl to a depth of about five centimeters. Shake the bowl from side to side to make the surface of the flour nice and smooth.
2. Hold the glass of water about a meter above the bowl and let just a few drops fall into the flour. You will see the drops will make little craters in the flour as they strike the surface. Notice how some of the flour gets blown out of the crater, forming a circle of debris around it?
3. Experiment with large and small drops until the whole surface of the flour is covered.

You have just made a model of the moon! Now compare your model to a close-up picture of the moon and see if there are any similarities.

30

Why Do Comets Have Tails?

Comets are sometimes called dirty snowballs because they're made of a lot of dust and dirt mixed in with ice. They're dark, almost as black as charcoal. They're also extremely old. They're leftover pieces from the huge cloud of gas and dust that formed our solar system—the extra bits that did not become planets. That makes comets time capsules, because they haven't changed for billions of years. They allow us to look back to a time before the Earth was even born.

So why are comets usually a surprise when they show up? Because most of the time they're invisible. We don't see them unless they have tails, and comets only have tails if they come close to the sun and warm up. Most comets hang out beyond Pluto, where it's cold and dark and they stay frozen. If they don't get warm, no tail. And no tail means the comet is too small to be seen in our night sky.

Comets cross the orbits of the planets, which means there is a good

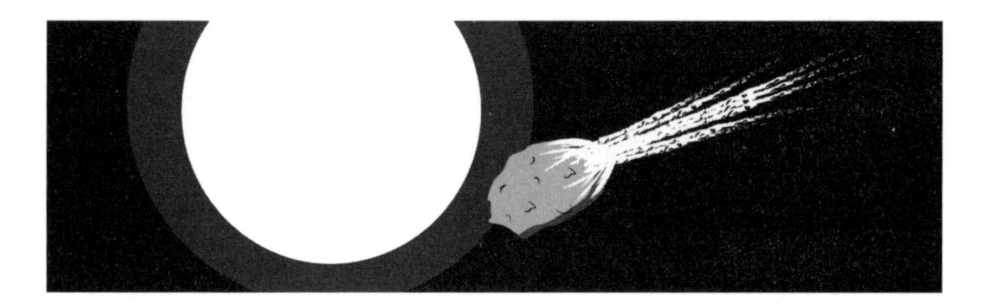

chance of a collision. In 1994, astronomers from around the world trained their telescopes to see, for the very first time, a comet running into a planet. And it wasn't just one collision. The comet, called Shoemaker-Levy 9, started out in one piece, but when it got close to Jupiter, the giant planet's powerful gravity tore the comet into pieces, forming a long train of comet pieces that plunged into the planet's atmosphere at speeds of more than two hundred thousand kilometers per hour. Gigantic explosions sent enormous balls of fire above the planet, and dark scars marked the face of Jupiter for weeks afterward. It was a spectacular event but also a little scary because if a comet can hit Jupiter, it means a comet could hit us. And we know Earth has been hit by comets many times in the past.

In 1908, a comet struck northern Russia. Fortunately, there were no people living in the area, but the impact flattened trees in the surrounding forest for hundreds of kilometers in all directions. If such a powerful event happened today near a city, it would be one thousand times more devastating than the first atomic bomb dropped on Hiroshima.

No one knows exactly how many comets are out there, but there are certainly more comets in space than there are people on Earth. Several robots have been sent to take pictures of comets. One robot, called *Stardust*, flew right through a comet tail and picked up pieces of the dust that were brought back to Earth for analysis. Another robot, *Deep Impact*, launched a heavy slug that smashed into a comet to see what was inside. As expected, the inside of the comet was made of a lot of water, ice, and dust, as well as some carbon compounds. And another robot, named *Rosetta*, followed a comet all the way around the sun and sent a lander down onto the surface of the comet itself. From these robotic adventures, we've learned that most comets are about five to ten kilometers across, which is the size of a small city.

If you could ride a comet, you'd get a tour of the entire solar system, passing by tiny, icy Pluto, followed by the giant blue-green planets Neptune and Uranus. You'd continue on past Saturn, with its magnificent rings. Lastly, you'd pass mighty Jupiter, the king of the planets. By that time, you'd start to feel warmer as you'd be closer to the sun. Then things would get really interesting. An amazing transformation would take place as your comet begins to warm up. Gases would erupt from below the surface in great geysers, shooting ice crystals into the sky. Snow would fall up! And those snow geysers would be seen from the Earth as a gorgeous comet tail. The funny thing is that pictures of comets make it seem like they're flying through space headfirst. While that's sometimes true, they can also fly tailfirst, meaning the tail sometimes wags the comet rather than the other way around.

A comet tail is like a flag that always blows in the direction of the wind. In space, there is a kind of wind of electrified particles that blows out from the sun in all directions called the solar wind. This is the force that always keeps the comet tail pointing away from the sun no matter what direction the comet is moving. As the comet comes in from the edge of the solar system and heads almost directly toward the sun, the tail is streaming out behind. But when it swings around the sun, the tail swings, too, always pointing away until the comet is flung back out into deep space tailfirst.

If you ever happen to see a comet, you are fortunate, because they are among the most beautiful of all celestial objects to cross our skies.

31

How Much Junk Is in Space?

We humans are messy creatures. We leave garbage all over the Earth and we also leave it in space. More than seven thousand satellites are in orbit around the Earth, but most of them are not working.

Satellites are machines and, like all machines, they wear out. That means there are a lot of dead satellites in space. In 2009, a dead Russian satellite called *Cosmos 2251* collided with an American communications satellite called *Iridium 33*. When these two objects hit, their two speeds together added up to more than forty-two thousand kilometers per hour. Needless to say, both satellites were completely blown to bits, and those bits added to the growing problem of space junk.

Dead satellites are only part of the space junk, though. Every time a satellite is launched into space, it rides on the back of a rocket, so there are used rocket boosters scattered in orbit. But there are also about a million

tiny bits of debris—burned-out boosters, nuts and bolts, even flecks of paint—circling the planet, all contributing to the increasing problem of garbage in space. In short, space is littered with potential hazards. The more objects we send into space, the greater the chances of them colliding with other objects, making more pieces of junk and more chances of collisions, which also create more junk . . . If this keeps up, there will be so much debris in space, no one will be able to go up there without getting hit.

All these pieces of space debris are scattered over an extremely wide area, but each one of them is moving incredibly fast—up to twenty-eight thousand kilometers per hour, which is faster than a bullet. A collision with a piece as small as your finger can have the explosive force of a grenade and could completely destroy a multimillion-dollar satellite, or worse, threaten the lives of astronauts in space.

Satellites are launched just about every week, but there's rarely a plan to dispose of them when they come to the end of their lives. Fortunately, some of them take care of themselves.

Objects in low Earth orbit, including space stations, tend to take care of themselves because they're not entirely out of the Earth's atmosphere. Even at three to four hundred kilometers above the Earth, there is still a tiny amount of atmosphere that objects fly through. That atmosphere drags on the satellites, causing them to slow down over time, so their orbits naturally decay until they plunge back into the thicker atmosphere below and burn up from air friction. In fact, sometimes what looks like a shooting star is actually a piece of space junk falling back to Earth. Even the International Space Station, four hundred kilometers above the surface of the planet, experiences a tiny bit of atmospheric drag, so the station has to be boosted back up on a regular basis. One day, it, too, will meet a fiery end in our atmosphere.

For the most part, smaller satellites will burn to nothing on reentry, so we don't have to worry about them too much. It's only when larger things—like space stations—burn up that some fragments of debris survive all the way to the ground. Of course, no one wants to be hit by a piece of space junk falling from the sky, but the problem comes when we try to predict exactly when and where the debris will fall.

A lot of space junk is moving so fast that it skips across the top of the atmosphere like a stone skipping across a pond. And like a stone, how far the pieces go and when they stop depends on a lot of factors: the shape of the stone, the angle it hits the water, and any waves that could get in the way. Some stones dig in immediately, while others seem to go on forever before coming to a stop and dropping straight down.

Dead satellites are odd-shaped and often tumble end over end, so how they will behave in the air is uncertain. The final impact point cannot be predicted with any accuracy until the last few orbits, and even then there is uncertainty.

That was what happened in 1979 with the fall of Skylab, the first American space station, the Russian space station Salyut 7 in 1991, and more recently, the Chinese space station Tiangong-1 in 2018. Scientists tried to reassure the public that there was little to fear by pointing out that the odds of being hit by debris falling from the sky are extremely low because very little of it actually reaches the ground. Also, most of the planet is ocean, so it is more likely that the pieces would fall in water than anywhere on land.

Tiangong-1 fell in the Pacific Ocean, but both Skylab and Salyut 7 overshot their predicted impact points. Bits of the American station fell down in Australia, and parts of the Russian station came down in Argentina. Thankfully, no one was hurt in these cases, but it shows how difficult it is to pin down the final resting place of an uncontrolled large object.

The best-case scenario was the intentional de-orbit of the Russian space station Mir in 2001. The 140-ton complex was the largest object to reenter the Earth's atmosphere, and it was controlled from the ground. Using a small rocket, the operators on the ground drove Mir toward the Pacific Ocean where the pieces fell harmlessly into the water. It makes you wonder how the International Space Station, by far the largest object ever to fly in space, will be brought down when its mission is over, sometime after 2024.

The rest of space junk in higher orbits is a much bigger problem. Because that junk is completely above the atmosphere, it will take thousands of years to drop to Earth, if it ever does. And with an ever-increasing number of satellite launches in the future, the amount of debris and derelict satellites will only increase as time goes on. If we don't do something about the problem, we could get to a point where it will become impossible to launch anything into space because of the hazard of running into junk.

And let's not forget the many tons of equipment that were left on the moon, such as lunar modules and rovers, boots, scientific equipment, and robot landers that no longer work. There are several dead satellites in orbit around the moon as well.

Even Mars has space junk already, with dead satellites in orbit around the red planet as well as dead robot landers and rovers on the surface.

So what can be done to clean up all of this space junk?

There are a range of ideas. The European Space Agency has a plan to send up a robot to capture derelict satellites, either with a mechanical arm or a net, and drag them down into the atmosphere. But that would require a huge number of robots.

Other organizations suggest sending up refueling robots that will act as space gas stations to fill the satellites' fuel tanks and keep them operating longer.

Other ideas are a little more radical. The Australians have proposed using a high-powered laser to blast the very small pieces out of orbit. The Chinese are working on a similar plan. But some people worry that a laser could be used as a weapon and weapons are not allowed in space . . . yet.

All of these ideas are very expensive because the amount of junk in space is huge and spread over such a vast area. Any cleanup method will take decades.

The best idea is to stop leaving junk up there in the first place. Already, we're designing rocket boosters that return to Earth for recycling. Future satellites should carry extra fuel that can be used to dispose of themselves at the end of their useful lifetimes.

There was a time when people didn't think twice about throwing garbage out a car window. Now littering is against the law. It is time for a similar law to be put into place for the entire international space community so we can clean up the dangerous litter that surrounds the planet and ensure we can safely travel through and explore the cosmos for many years to come.

YOU TRY IT!

Space Junk Food

The biggest problem with trying to clean up space junk is that there is so much of it spread over a massive area. Each one of the many pieces of junk are circling the Earth in a different orbit and at a different distance out in space, so it is hard to get to them all.

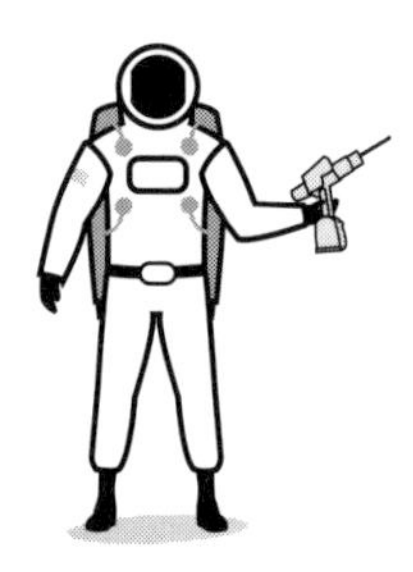

WHAT YOU NEED

1. An orange, apple, or a baseball (any small ball will do)
2. Some sugar

WHAT TO DO

1. Place the ball on a table and sprinkle some sugar in a circle around it, out to about the width of your hand.
2. Wet the tip of your finger with your tongue and try picking up the sugar by dabbing your finger on the table.
3. Rub the sugar off your finger and repeat. How many dabs does it take to capture all the sugar?

The ball at the center of the table is the Earth, while the sugar around it represents all of the junk circling our planet. Each dab of your finger mimics a spacecraft sent into orbit to clean up the junk. You can see that only a little bit of junk gets picked up each time, so it will take hundreds, if not thousands, of missions to clean it all up. Perhaps a better idea is to make sure that missions in the future don't produce any junk at all!

CONCLUSION

The Wonder of It All

The universe is an unbelievably huge, infinitely interesting place that we have only begun to explore. What we see with our eyes is only a tiny part of what is really out there. We have discovered that we live on a ball that is spinning at dizzying speeds through an ever-expanding space that includes other fascinating worlds, swirling galaxies, and mysterious black holes.

In the future, robots will visit more worlds in our own system, drilling through the ice of Europa to search for life in its salty ocean, or sailing the methane seas of Titan. Human explorers will return to the moon and then head out to the dusty red soil of Mars, searching for clues to any life that might be there now or fossils of creatures that lived there long ago. And speaking of life, perhaps one day soon we will make contact with another civilization, aliens from another world, and find out who our neighbors in space really are.

As our telescopes get bigger, more and more of the universe comes into view. And that raises more questions, such as: What went "bang" at the beginning of the universe? Was there anything before that? Are there other universes? How does life get started on a planet? Why can't we go faster than the speed of light? Is time travel possible?

Who knows what we will discover as we seek out the answers to those questions.

While we look farther and farther out into space, we also think about our own place in it and on our home world. Everyone who has traveled in space has marveled at the beauty of our blue planet when seen from the outside. And while other planets are very interesting to explore, none of them are like our own. None of them have air that we can breathe, a climate warm enough so you can step outside in shorts and a T-shirt, lakes to swim in, or flowers to smell. In fact, every planet we know of will kill you. As far as we know, Earth is the only place where you don't have to wear a space suit just to take a walk.

So studying the wonders of the universe is also a study of our home planet, a beautiful blue oasis of life floating in a fascinating but deadly sea of darkness. It may be a small planet, but it's all we have, and that makes it worth taking care of.

ACKNOWLEDGMENTS

I would like to thank Nita Pronovost and Brendan May for their patience and skill in editing the manuscript, John Pearce for bringing my ideas to publishers, and Jennifer Hartley for her support and encouragement.

Photo by Jennifer Hartley

BOB McDONALD has been the host of CBC Radio's *Quirks & Quarks* since 1992 and has worked in TV and radio for more than forty years. He is a regular science commentator on CBC News Network and science correspondent for CBC TV's *The National*. He has been honored with the 2001 Michael Smith Award for Science Promotion from the Natural Sciences and Engineering Research Council of Canada, the 2002 Sandford Fleming Medal from the Royal Canadian Institute, and the 2005 McNeil Medal for the Public Awareness of Science from the Royal Society of Canada. In November 2011, he was made an Officer of the Order of Canada. In 2014, an asteroid designated 2006 XN67 was officially named BOBMCDONALD in his honor. Bob lives in Victoria, British Columbia. Visit him on Twitter @CBCQuirks.